Simulation in Media and Culture

Simulation in Media and Culture

Believing the Hype

Edited by Robin DeRosa

LEXINGTON BOOKS
Lanham • Boulder • New York • Toronto • Plymouth, UK

Published by Lexington Books
A wholly owned subsidiary of The Rowman & Littlefield Publishing Group, Inc.
4501 Forbes Boulevard, Suite 200, Lanham, Maryland 20706
www.rowman.com

10 Thornbury Road, Plymouth PL6 7PP, United Kingdom

First paperback edition 2013

British Library Cataloguing in Publication Information Available

Library of Congress Cataloging-in-Publication Data

The hardback edition of this book was previously cataloged by the Library of Congress as follows:

Simulation in media and culture : believing the hype / edited by Robin DeRosa.
p. cm.
Includes bibliographical references and index.
1. Mass media—Social aspects. 2. Mass media and culture. 3. Reality. 4. Virtual reality.
5. Simulation methods. I. DeRosa, Robin.
HM1206.S563 2011
302.2301—dc23
2011021140

ISBN 978-0-7391-6477-8 (cloth : alk. paper)
ISBN 978-0-7391-8458-5 (pbk. : alk. paper)

Contents

Preface

In May of 2003, I was at home in New Hampshire waiting to give birth to a baby and generally watching too much television. I vividly remember the local news breaking in to the travel program I was enjoying to deliver some traumatic information. The Old Man of the Mountain—a natural granite profile that sat atop Cannon Mountain and was New Hampshire's much beloved state symbol—had separated from its perch, slid down 1,200 feet, and landed in a pile of rubble. It was discovered by two early morning hikers, who marveled at the debris, glanced upwards, and got the shock of a lifetime as they realized what the bare mountaintop indicated. As a recent transplant from Massachusetts, I think my biggest reaction to the news was minor annoyance that someone in as swollen a state of pregnancy as I was in should be deprived of the rest of her travel program over such incidental happenings. But as I watched the coverage, my interest grew. Locals were gathering by the base of the mountain to lay flowers and wreaths. Residents being interviewed by the news crew fought back tears as they remembered the Old Man. Though nobody (except the Old Man) had been injured in the slide, the news coverage resembled the kind of coverage one sees in the wake of a national tragedy in which innocent people have been killed and people's senses of security have been dramatically shaken.

Immediately after the fall, local politicians and businesspeople began trying to spin the event in such a way so as to preserve the hope that tourists would continue to flock to New Hampshire even after the loss of one of our most popular attractions. Jayne O'Connor, a paid promoter for the White Mountains, put it this way: "As far as the Old Man goes, we now sort of have our own Mayflower, or one of those attractions that are long gone, but are remembered and people still visit the sites."[1] How does one not just survive, but *thrive,* in the wake of loss? This was, for my new home state, not just an economic question, but also an emotional and cultural one. Could New Hampshire reconstitute a tourist site and a state symbol in the rubble of what had been? As I watched the Old Man site journey from a concrete—indeed, a granite—formation to an *absence,* I noticed how the original granite properties of the Old Man seemed to have been overshadowed from the beginning by the symbolic properties of the site. In the wake of the loss of the Old Man, the original Old Man was gradually revealed to have been constituted all along by a complex series of performances and simulations.

To understand how this has been the case, one needs to begin with the question of the "original" itself. A sighting of the Old Man was first recorded in 1805, when a road crew near Franconia Notch made a note of the profile. There is no evidence that anyone saw the Old Man before this date, though many legends claim that the Old Man has been around since the Ice Age, that he was worshipped by Native Americans as long as they have lived in the Northeast, that white explorers encountered him as far back as the 1650s. In fact, most literary and historical accounts of the Old Man from the nineteenth century echo resident Lucy Crawford's 1846 account: "The storms of ages seem not to disturb the resting place of this stern old man, or cause a wrinkle to be furrowed on his brow."[2] Despite lore that suggests that the Old Man is as eternal as rock and air itself, geologists now believe that the Old Man—the angles and outcroppings that comprised the vivid rock profile—were probably not visible until about 1805, because of the rate of erosion and exfoliation of the talus slope on Cannon. This would certainly explain the lack of earlier recorded sightings. Of course, when this research emerged about six months after the Old Man's fall, locals were understandably riled up. The eternal and enduring nature of the rock face was key to its symbolic value for the rugged state of New Hampshire. The "Old" Man was never actually old. That he symbolized the eternal agelessness of nature was due not to "reality," but due to the narratives that were constructed about his (mythic) origins. Because these origins are mythic, they are also pliable, and they can shift to accommodate newly emerging events and research; spin doctors interceded again. Art Historian Robert McGrath of Dartmouth explained why the wards of the "Legacy Fund," a campaign to memorialize the Old Man, joined with the spokesman for the New Hampshire Cultural Commission in downplaying both the Old Man's fall and the research that suggested he was not very old at all; the Old Man, argued McGrath, "is in the national imaginary—in license plates, highway signs, paintings, literature. That's more powerful than the reality."[3]

Despite the fact that the Old Man stood for the natural beauty of the state, and for the authentic ruggedness of our terrain, what most people don't realize is that he was held together by a complex system of turnbuckles, epoxy, and industrial bolts. Just sixty-seven years after the profile was discovered, the Appalachian Mountain Club noticed that the forehead was slipping, and work began soon after that to shore him up. By the time the Old Man was a bonafide tourist site, he was no longer a natural phenomenon, but this had little effect on his ability to stand for nature unspoiled. As Monica Davey of the New York Times claimed, "The Old Man was no sculptured, dolled-up Mount Rushmore."[4] At least, not symbolically. When the Old Man fell, the mountaintop was not quite bare after all: the elaborate fastening system, broken and bent, reached outward into the air. Today, tourists can pull off at the former "Old Man Viewing Site," and look through a viewfinder to see the twisted metal at the top of the slope.

In 2009, I took my daughter—born so soon after the Old Man's fall—to the Profile Awards ceremony, which recognizes New Hampshire residents who, like the Old Man, personify the spirit of the Granite State. Even six years after his fall, the Old Man still elicited tears that day from his caretakers, who spoke of the loving way that they had applied glue and bolted rock over the years. Plans were unveiled for several memorials, all of which remain unbuilt at this time. In fact, one plan won the prestigious AIA Merit Award for Unbuilt Architecture for a memorial that would allow viewers to stand inside a glass replica of the Old Man and look out onto the valley below, in a gesture that acknowledges that the Old Man himself exists less in reality than he does in memory. But in a way, the Old Man was always already a memorial to himself. Invented as much rhetorically as he was by glacial activity, the Old Man's history was constructed in a backwards movement from the present. His rugged, natural reality was created semiotically through art, literature, souvenirs, and tourist signage. For the Old Man, loss and absence are not just about his 2003 fall, but central to his past and his future.

The simulation, then, will not just come when at last the final memorial is unveiled at the former site of the Old Man of the Mountain. The simulation began when the Old Man was founded, and his ancient and steadfast (if unreal) past began to emerge through narratives. By the time the bolts were being placed, the simulation was being submerged, subordinated to a new invented reality about the profile's natural authenticity. Currently, in the stage of the unbuilt memorial, the state gathers around its loss, invoking nostalgia for a time when the wilderness was truly, if ironically, wild. The winning memorial design, the one that will supposedly be built one day, depicts a series of massive granite slabs that, when lined up properly, will come together to make the outline of the Old Man for each viewer's eye. This fragmented Old Man, visible only when the viewer is seeing properly, will become its own simulacrum, a total replacement of the real that once (perhaps) existed, and a more engaging, controllable, and enduring tourist destination than the former Old Man could ever have been. In a short essay about the Old Man published shortly before his fall, Peter Warshall provocatively argues that the Old Man's hyperreal qualities are exactly what generate the sacredness of the site for tourists and for citizens of the Granite State[5]; this may be even more true now that the Old Man has slid into physical absence.

Scholars familiar with Jean Baudrillard's work will recognize in the progression of the Old Man's symbolic history a nearly perfect enacting of his stages of the simulacra. The relationship of the simulation to reality shifts as the simulacra develops: at first, it reflects reality; next it hides it; then it hides that it is hiding it; and finally, it replaces it so completely that the original is impossible to decipher. What the simulacra yields then is a reality that appears more real than the real itself: a *hyperreal.*[6] This collection draws centrally from Baudrillard's theory of the simulacra, and the essays here describe how hyperreality functions in media and culture. From my office here in New Hampshire, I enjoy a fond—if odd—nostalgia for the new/lost Old Man of my adopted state, and he has presided with both dignity and irony over my work on this collection.

~R.D.
Plymouth, New Hampshire

Notes

1. Holly Ramer, "Remembering the Old Man of the Mountain," *The Concord Monitor,* May 28, 2008.
2. Lucy Crawford, *History of the White Mountains* (Hanover: Dartmouth Publications, 1966), 198.
3. Monica Davey, "For the (Very) Old Man, Is a Facelift Unseemly?" *The New York Times,* May 7, 2003, 22.
4. Ibid., 22.
5. Peter Warshall, "Wilderness and the Hyperreal," *Whole Earth* 104 (Spring 2001): 54-5.
6. Baudrillard, Jean. *Simulacra and Simulation* (Ann Arbor: U Michigan P, 1994).

Acknowledgments

This book is indebted to two wonderful panels that I chaired at recent conferences of the Northeast Modern Language Association; to participants in "S(t)imulated Realities" and "Postmodern Tourism," I offer many thanks for your lively dialogue and energizing ideas. My students in "The 'Real' World," an advanced topics course in English here at Plymouth State University, also asked many fruitful questions and explored many avenues with me as I was beginning to conceive of this project.

I am very grateful to Plymouth State for the sabbatical that allowed me the time to compile and review this work. And to the contributors, I thank you for the high quality of thinking and writing that you have done here, and for the new professional friendships that I hope will endure now that the project is complete.

I am grateful to two individual people who were instrumental in helping me bring this collection to press. My editor at Lexington Books, A. Lenore Lautigar, has been supportive and responsive at every step of the process. And most significantly, my student and editorial assistant, Lisa A. Riley, without whom this collection never could have been realized. She is all that one could hope for in a collaborator: organized, positive, diligent, and brilliant.

Introduction

Robin DeRosa

Taking in just under three billion dollars so far, James Cameron's 2009 film, *Avatar,* is currently the highest grossing movie of all time. The film follows the story of a young soldier, Jake Sully, whose recently-deceased twin brother had been immersed in a cutting-edge operation involving a custom-made avatar. As Jake learned to pilot his brother's avatar body, Cameron garnered massive attention for the special effects that integrated live human actors with complex digital animations. The content of the film spins around an axis of simulation: an avatar of a twin becomes more authentic than either Jake's real self or the ultimately-forgotten original brother; and the form of the film, which prevents viewers from distinguishing between "real actors" and green-screen projections, contributes to the dizzying slippage between a real world and a constructed one. As *Avatar* debuted, Tina Fey captivated America with her televised Sarah Palin impersonations; Palin responded by telling *New York Magazine* that she wanted to go on *Saturday Night Live* to spoof Fey's own American Express commercials. The last ten years have seen an explosion of simulated realities in everything from film and television to online gaming, technology, politics, tourism, and social identities. Informed by a theoretical framework that draws from Jean Baudrillard's seminal 1985 text *Simulacra and Simulation*, as well as from work by Umberto Eco, Homi Bhabha, and Plato, this collection explores how simulated realities—especially those that become "more real than the real"—function in recent examples of media and culture. Together, the contributors posit innovative ways to discern meaning in a world where reality is always mediated, and where representations replace authenticity.

It is, of course, television where hyperreality got its strongest initial foothold in our cultural imagination. But reality television has evolved past the first decade, where prize-oriented shows such as *Survivor, Fear Factor,* and *Big*

Brother ruled the airwaves. The essays in this collection look at a new generation of reality-infused television programs which represent a shift and expansion of the genre. In "Famous for Being Famous," Jenna Gerds examines a newer type of reality TV star: the celebrity who maintains her fame by being herself, though in this case, the "self" becomes an elaborate set of exterior projections of an interior "reality" that may or may not even exist. Unlike actors with a skill to market or reality TV stars of an earlier generation who gained fame by competing against each other, these celebrities—from the "real" housewives to the socialite we all love to hate, Paris Hilton—glamourize the superficial and both subvert and reinvigorate the real itself as their daily lives are televised and used by viewers to shore up an understanding of their own "really real" lives.

In "Simulating Dinner," Glenda Shaw-Garlock expands the reality television genre even more by reconsidering the cooking show. Though she traces the history of televised cooking programs back to their earliest days, she redefines the genre from inside the parameters of contemporary reality TV. Like Gerds, who is interested in how TV audiences and blog readers interact with celebrities' lives, Shaw-Garlock considers how cooking show spectators relate to the personas of celebrity chefs. Ultimately, she suggests a new paradigm for thinking about the world of the televised chef, a world in which *watching* dinner replaces eating it. A new nuclear family emerges, with new domestic traditions and newly imagined relationships to the kinds of comforts and pleasures that intimate family connections engender. These connections, situated halfway between a Las Vegas restaurant and a cable television station, make the original real family dinner fade into a kind of lackluster grayscale that pales next to the luscious Nigella Lawson and the BAM of Emeril Lagasse.

This kind of Baudrillardian simulacra, in which the simulation ends up eclipsing the original from which it presumably originated, is also apparent in Naomi L. Fosher's essay on *Seinfeld* and *Curb Your Enthusiasm,* "Imitation Meets Simulation." By examining the slippages between "Larry Davids" (Larry David the *Curb* star, Larry David the creator of *Seinfeld,* George Costanza the simulation of Larry David, etc.) Fosher delves into the porous boundary land between television sitcom and reality show. By looking at the ways in which one sitcom bleeds into another, and how both bleed off the pages of their scripts into a world that is supposedly "real," Fosher delineates and blurs the line between the real and the copy. In her discussion of how editing and improvisation interact on the set of *Curb* and how actors interact with their characters inside and out of the programs, she demonstrates how the simulacra ultimately destabilizes the distinct outline of the real itself.

While *Curb* reinforces its own status as a "real" show by playing against *Seinfeld's* status as fiction, the American TV drama *Supernatural* does a similar thing by playing off its own invented versions of reality. As Michael Fuchs demonstrates in "Trapped in TV Land," the show develops its simulacra-related themes as it plays with live-action role-playing (LARPing), meta-fiction, and even the hyperreal qualities of a world where death is a rebirth into another shade of life. Ultimately, Fuchs is interested in how the show's postmodern play can create a kind of real—if ironic—emotional response. Linking theories of the simulacra to Umberto Eco's theories of narrativity, Fuchs explores how *Supernatural* makes sense of the real, human world through storytelling, role-playing, imitation, and replication.

If the essays in the first section of the collection expand the definition of reality television, the essays in the second section aim to move past the first generation of simulacra-related films and film-theory. Most critical response to hyperreality in film has focused on movies that have specifically played with the intersection between the real and hyperreal worlds: films such as *The Matrix* and *The Truman Show*. In this collection, the four essays that focus on film push into less obvious examples of the simulacra. In "The Gentleman's Gentleman," Ann K. McClellan looks at the role of the British butler in literature, television, and movies, with a special emphasis on the 1993 film, *The Remains of the Day*. In her essay, McClellan mines the complex relationships among commodification and exchange, class position, postmodernism, and the construction of Britishness as they interact through the similarities and differences between Mr. Stevens and his master, Lord Darlington. While McClellan focuses on the instability of class and national power in "England" (both the "real," historical England and the England of American narrative), Randy Laist moves to the instability of the changing American cultural attitude of the 1980s and 1990s. Laist takes us through the transition from Arnold Schwarzenegger's earlier films, which make an anti-postmodern statement (as Laist puts it, "that reality is real and it will kick your ass"), to his next decade of films, which curve toward irony and self-consciousness. In "The Soft Bodies of Arnold Schwarzenegger," Laist characterizes this second filmic era as "90's hyperreal cinema," and examines how Schwarzenegger's films of this period (for example, *Total Recall* and *Terminator 2)* blur boundaries between human and nonhuman, confound temporal rules, and cause viewers to reassess our understanding of the relationship between the real and the represented.

Amarnath Amarasingam's essay, "Jesus in Hyperreality," looks at Mel Gibson's 2004 film *The Passion of the Christ,* and asks questions about how films based on scripture affect lived religious experience and the interpretation of scripture itself. As viewers of the graphic film "bear witness" to the suffering of the character of Christ, the simulation of Christ eclipses the "real" Christ, visible now only through the multiple lenses of the film's research background and directorial vision. Amanda LeBlanc summarizes the trope evident in each of these film-related essays as she examines the recent coalescing of what I might call "avatar culture." Starting with Facebook and ending with James Cameron's 2009 blockbuster, *Avatar,* LeBlanc's essay, "Making Love, Making Violence Avatar-Style," theorizes the avatar as occupying a liminal, "hymeneal" position. Following Derrida, LeBlanc posits a complex way of understanding the avatar, one that sees a simulation as less of a contrast with the real than an intimate partner to it. After looking back to *The Matrix* and other seminal hyperreal texts, LeBlanc extends Laist's characterization of the simulacra in cinema through the current moment, and revisits themes of class raised by McClellan and love for and faith in the simulacra raised by Amarasingam.

In Part III, we turn our attention to games, already a mirror of the real in terms of their very definition. The games addressed in this section are technology-related, and the authors work to theorize connections between gameplay, reality, and the mediated space of the delivery system. In "Planting Crops in the Hyperreal," Emily Hall draws from Slavoj Zizek to consider how the real world of work intersects with the simulated work environment of Facebook's wildly popular game, *Farmville*. Informed—as many of the essays in the collection are—not just by a postmodern philosophy, but also by a Marxist critique of the fetishization of the commodity—Hall explores why people would give up real time and real money to earn the simulated rewards offered for tending their online farms. In "Fear and Loathing in Second Life," Cathie LeBlanc is similarly interested in the overlaps between the real world and the online world, though in this case the online world is Second Life, a virtual community that many hoped would become a utopic space where members could shed their real-world identities and live a life of complete freedom from bodily regulation. LeBlanc argues, however, that the medicalized gaze, theorized by Foucault and popularized by television shows like *Extreme Makeover,* follows Second Life users as they design their avatars from within the constraints of the same internalized, disciplinary surveillance that regulates real-world bodies in Western consumerist culture.

In her essay, "Total Immersion and the Total Screen," Katheryn Wright asks a more form-related question, considering how screens function in the simulated world of games. Through a discussion of particular video games such as *Dead Space* and *Tom Clancy's Splinter Cell: Conviction,* Wright's essay examines how these games use a series of screens to navigate the simulated worlds; she uses examples to revise Baudrillard's concept of the "total screen," in which there is a collapse of the distance between game and player. For Wright, the screen's *materiality* (rather than its disappearance) works to construct an (ironic) real, a "screened reality" of sorts. If Emily Hall and Cathie LeBlanc identify the dangers that accompany the blurring of real and screened worlds, Wright begins to deconstruct what she calls the "deep-seeded fear of change in the name of conformity" that the Baudrillardian total screen engenders. In her essay, "Seriously Fun," Lian Amaris offers a similar corrective to Baudrillard. Amaris argues that ARGs (alternate reality games) like the one that accompanied the release of Christopher Nolan's 2008 film *The Dark Knight,* which combine multiple forms of interactive communication, return the players to the public, social, sphere after the era of the isolated PC-user. She suggests that ARGs can resist charges of nihilism launched against (and sometimes by) postmodern philosophers because meaning gets produced through the experience, active communication, and social interaction of the games.

The critique of theoretical approaches to simulation continues in Benjamin D. Hagen's "Transgressive Simulation," in which he argues that critical writing on wrestling has not kept pace with either theory in general or with wrestling itself. Hagen revises Barthes' comments from his 1957 essay, "World of Wrestling," considering how earlier understandings of wrestling as a purely intelligible space in which the dividing line between the simulation and the real is clear to all involved is challenged by newer forms of the game, such as the underground promotion, Extreme Championship Wrestling (ECW). By looking specifically at violence in ECW, Hagen posits a new way to think about the dissonance between real pain and the scripted aspects of the show. Concerned not just with wrestling, Hagen's argument extends to the broader issue: rethinking "the ethics of entertainment" in a postmodern world. Similarly broad in its repercussions is Mihaela P Harper and Andrew J Ploeg's essay, "Drafting the Hyperreal," which considers the role of agency in the hyperreal world of fantasy sports. Departing once again from Baudrillard, this essay sees the hyperreal space not as one that drains subjectivity from players, but as one that deconstructs the binary between between real and fantasy. A kind of dis-identification accompanies the shifting of players' allegiances between "real" teams and players and "fantasy" teams and players, and this in-between position, non-dichotomized and paradoxical, offers new ways of theorizing the hyperreal itself, and the meaning that is possible in its territories. All of these essays probe the central questions: how is the game separate from and/or indistinguishable from the real world? what are the real repercussions for real people who play games? and how is meaning made in spaces where lines between real and simulated are troublingly and/or liberatingly blurred?

In the fourth section, we shift out of our genre-focused paradigm, and instead look at how culture, nation, state, and city are constructed in a hyperreal world. By interrogating the formation of identity categories and how they interact with history and semiotics, the authors in this section offer new paradigms for thinking about who we are and the places we inhabit. In "Room(s) for Sight," Matthew Cornish uses Rimini Protokoll, a German interactive theater company who created the movement known as "Reality Trend," to interrogate the relationships among globalism, capitalism, intimacy, and artifice. In an examination of two theater productions, *Wallenstein* and *Call Cutta in a Box,* Cornish explores tensions between (for example) performers and audience members, Europe and postcolonial India, and representation and life, and posits a new way of thinking about a human's relationship to the hyperreal, a relationship that, while wrapped in the compromise of simulation, can be sensory, fulfilling, illuminating.

Brian Johnson is interested in "Nazis and Nazi Costumes," but his essay is fundamentally a plea for a new ethics of hyperreality, a theory which would allow us to rethink the "axis of (good and) evil" in a simulated landscape.

Johnson deconstructs the idea of Nazism as a transcendent signified, a stable indicator of evil, and instead argues that Nazism ultimately comes unfixed and unhinged from the real as it moves from historical event to cultural sign. Provocatively, Johnson argues for a kind of hyperreal Holocaust Studies, one that would be equipped to weigh the significance of Nazi signs rather than Nazi history; by insisting on the value of ethics to the study of hyperreality, Johnson ironically questions the strength of fascism as a comparative barometer for all ethical questions.

The next two essays, David de Posada's "Operational (Hyper)realities in the Exilic Labyrinth" and Leslie Haynsworth's "Having it Both Ways," consider, in different ways, the meaning of "place" in postmodern culture as described by two contemporary novelists. De Posada looks at the work of Cuban-American writer Roberto G. Fernández, whose Cuban characters are wrapped in a multi-layered kind of escapism as they move to a new country in search of the spectral American dream. De Posada defines a relationship between nostalgia and fantasy, which engenders an exilic identity that is both remembered and invented, authentic and hyperreal. Haynsworth's essay on British writer Geoff Nicholson's novel *Bleeding London* asks questions about whether we can both value the real *and* appreciate the control that we have over simulations of the real—simultaneously. By using Benjamin's concept of the "aura," Haynsworth rethinks the "real" in postmodern terms, exploring how subjectivity and perspective mediate our direct experience of any original. She explores how tourist cities—in this case, both the real and the fictionalized London—exist on a continuum of simulation, and how the tourist gaze can be both awe-inspired and authoritative.

The final essay in the collection, Lisa K. Perdigao's "There's really nothing left to explore," takes as its subject the field of Florida Studies (or, as it might sometimes be conflated, Disney Studies). She argues that Florida Studies has become its own subfield within Cultural Studies, and she works to reimagine it as a simulacrum of itself. By positioning the master narrative of Florida's history against its simulated, Disneyfied history (or "Distory"), Perdigao remaps Florida for the twenty-first century, positing the state—now located in a postmodern geography not necessarily (or not merely) tied to its physical site—as a land of paradoxes; as utopia and dystopia collide, and origins slide to a profound but inaccessible space beyond the map, "Florida" becomes a contradictory, but ultimately open and infinite, new terrain.

Together, the essays in this collection work to extend Baudrillard's concept of the simulacrum into a new era. They revise and question older theoretical approaches to the study of simulation, and they use examples from media and culture to expand our understanding of what it means to live a real life in a hyperreal world.

Part I

The Small Screen: Reconsidering Reality TV

Chapter 1

Famous for Being Famous: Celebrity Socialites and the Framework of Fame

Jenna Gerds

Before I was shot, I always thought that I was more half-there than all-there—I always suspected that I was watching TV instead of living life. People sometimes say the way things happen in movies is unreal, but actually it's the way things happen to you in life that's unreal. The movies make emotions look so strong and real, whereas when things really do happen to you, it's like watching television—you don't feel anything. Right when I was being shot and ever since, I knew that I was watching television.

–Andy Warhol[1]

Several years ago, my father—who aspires to be the "heppest" man of his generation—pointed to an image on the TV screen and asked me, "Who is that?" I responded, "Paris Hilton, Dad." Then he asked, "What does she do?" and I told him, "She's famous for being famous." This phrase, overused and underexamined at the time, is the subject I would like to take up in this essay. How does one become famous for being famous? What does that imply? Perhaps that a celebrity was born famous or became famous accidentally or through no fault of her own, and then their fame increased from that point.[2] Or perhaps something more complicated: that there is no real collective recollection of why or how that person became famous, but the fame seems to repeat itself, a seemingly endless copy of celebrity with no real origin, until one day, unnoticed and unlamented, her time in the spotlight expires.

The celebrity socialite is a subject I've chosen specifically for what she is not: namely, overtly and recognizably talented. Thus, she epitomizes the concept of being famous for being famous because the reasons she has catapulted into the limelight are particularly elusive. As a result, celebrity socialites rely upon a number of modes or tactics that signify celebrity in the hopes of crystallizing a fleeting, ephemeral fame into a longer lasting, more concrete kind of fame now often associated with an older era of Hollywood stardom. Scholars have categorized signifiers of celebrity in various ways, but "reality" stars that are famous for living their lives in the public eye seem to rely heavily upon just a select few of them: 1) an emphasis on the possession and circulation of capital, 2) an artificial intimacy with their audience—or the blending of public and private, and 3) a superficial quality that masks or obliterates the differences between real and false. Advances in technology have helped to facilitate the socialites' transformation into celebrities; the advent and ever-increasing capacity of the Internet are crucial to this metamorphosis, as are Internet facilitated platforms of expression like blogs, Facebook and Twitter. Another critical development has been the evolving importance of cell phones: omnipresent, Internet-ready and equipped to multi-task. Paradoxically, it seems that while not all fans of these women are able to openly recognize and acknowledge the celebrities' connection to a few of these categories of signification, many of them do. This acknowledgment can cause fans to either love or hate a socialite; strong sentiments such as these stave off public indifference and comprise a fourth mode of celebrity. Regardless of whether it is admiration or disgust that such acknowledgement fosters, fans remain interested in what the celebrity does anyway, despite the fans' ability to lift the veil of hyperreality. Indifference is the death knell for a celebrity socialite, and when one ceases to produce a reaction it signals the approach of the end of her fifteen minutes. Thus, fans that love to hate the socialite are just as important as the ones who simply adore her.

A paradox exists regarding celebrity socialites in that their names and images seem to be on every screen and page, on the lips of so many Americans, yet not many people actually seem to *like* these women. How, then, do they manage to get and stay in the spotlight, no matter how brief it may be? This essay does not analyze the glossy images in magazines or the grainy night vision of "accidentally" released sex tapes. Rather, in an attempt to uncover why the public is drawn to this type of celebrity, I've chosen a media reception approach that takes into account editorials, blog posts and comments posted on the Internet, as "real" or as unmediated an expression of public sentiment I could find. And I've found that reality television celebrities seem to be attached to the modes of culture and commodity that movie stars have been linked to for decades. The difference between a celebrity socialite and an actress is that the socialite uses these tactics in an increasingly exaggerated and intensified way. In fact, by necessity she clings closer to the ephemera that signify stardom to retain an audience, in place of being admired for any kind of overt performance or demonstrable talent such as acting or singing. This does not mean that she doesn't act, only that acting becomes almost indistinguishable from not acting and that her career did not begin as someone who plays a role on screen. Instead, every public appearance becomes a performance, every sound bite or video clip seemingly captured by accident is a press release. This type of celebrity has truly become famous for being herself, or at least a version or caricature of herself because a "real self" is essentially unidentifiable.

And *she* does seem to be primarily female, as well as white, although there are exceptions to both. She possesses a large of amount of capital that she does not seem to have earned through means of her own initially, and she spends that capital at will. She might parley her money into other investments, the way that she might parley her celebrity as a social being into other opportunities to gain fame or capital. Reality TV has given the socialite, who has always held some kind of public interest, a wide-open forum to display her uniquely nebulous capabilities. Then, in turn, being on a reality television show in some ways cements her status as a celebrity because she is now exposed to a large audience through a medium that is capable of creating stars in the eyes of the public.

Artificial Intimacy—Public vs. Private

The ability of certain socialites to pierce through the stratum of "ordinary person" to achieve some level of celebrity status is not a new phenomenon. Although the appearance of society pages that reported scandal and gossip among the upper class appeared in newspapers during the nineteenth century, an article by Robert E. Park published in 1923 delineating the history of newspapers suggests that the "first newspapers were written or printed letters; newsletters they were called. In the seventeenth century English country gentlemen used to employ correspondents to write them once a week from London the gossip of the court and of the town."[3] Moreover, Park believed that "[t]he first newspapers were simply devices for organizing gossip and that, to a greater or less extent, they have remained."[4] It was not until the twentieth century that the actual word "socialite," a term that hinges on class status, was used. A 1940 essay by Joseph J. Firebaugh in the journal *American Speech* notes that it was *Time* magazine that first coined the word in 1929. Firebaugh claims it "refers to a member of that class of people known through courtesy as the upper class, or high society, and is formed from the two words *social* and *light*."[5] Although the celebrity socialite is not always as polite and refined as this definition connotes, she does tend to treat her fans with a certain sense of decorum. Celebrities of all kinds know the importance of making one's fan base feel appreciated. Those that do not take time out to sign autographs are criticized for it, and those that do special things for fans—go to their birthday parties, visit them in the hospital, rally around their charitable causes—are praised. But reality television stars that are famous for reasons not immediately ascertainable seem to realize the tenuousness of their fame and make an even greater attempt to maintain a connection to their fan base.

By actively cultivating a fan base through publicizing private images, celebrity socialites are usurping a technique pioneered long ago. In his essay, "Celebrity Watching," Michael Newbury comments on Herbert G. Goldman's book, *Banjo Eyes: Eddie Cantor and the Birth of Modern Stardom*. Goldman credits vaudevillian actor Cantor with being one of the first stars to "foster the illusion of intimacy between himself and an anonymous audience, a kind of mass-mediated closeness that would become characteristic of modern renown."[6] Cantor's private life was made public and this allowed his audience to feel like they knew the actor personally. Because of this perceived connection, Cantor's audience then followed him as he switched media to radio and eventually television. But, just as the popularity of newsworthy socialites precedes the twentieth century, so does the relationship between star and audience. Newbury notes that "marital infidelities, courtships, purchases and pastimes of Broadway stars" appeared in the news before Cantor, even in the nineteenth century.[7] "Well before Cantor, the private and public lives of stars had been interwoven, and not so much by any individual performer as by the restructuring of the newspaper and theater industries, their shared interests in providing material and publicity for one another."[8] The symbiotic relationship of stars and tabloids is clearly ever-present today, as well, facilitated by the structures of reality television and the Internet.[9]

The American public reads all about the breakups and makeups of stars, and a change in relationship status is one of the private moments that receives the most public attention. There are also a multitude of television programs that show us what celebrity homes look like—including tours of their bedrooms and the insides of their refrigerators, what kind of cars they drive, what kind of gifts they buy their babies, and how much it all costs. With this kind of detailed information, it might be difficult to feel like you *don't* know something intimate about a celebrity. Marilyn Monroe's nude calendar photos may have appeared in the first issue of *Playboy*, but reality stars of today take this literal shared intimacy to another level. The middle of the first decade of the twenty-first century was marked by a rash of young female celebrities filmed and photographed partially or completely naked. The bodies of Lindsay Lohan, Paris Hilton, Britney Spears, Kim Kardashian, Tara Reid and several others were all intentionally or unintentionally exposed on film. And once you've seen a celebrity's cesarean section or breast implant scar, the boundary between staged intimacy and real intimacy (*too* intimate for many viewers) becomes blurred and confusing. In many cases, you might actually have to actively work to avoid seeing nude photos of a star. They are displayed on reputable news stations, censored by a meager black bar in the requisite place. They are sent to us in emails by friends and coworkers. Perhaps this overexposure is linked to their most frequent medium—reality television. Even if the exposure is staged and the breasts are implants, those are real naked bodies on the screen. The female body of the celebrity socialite is public property in practically every way, and this overexposure is one of the means by which she garners public attention.[10]

Seeing and hearing about a celebrity's sexual liaisons and multiple relationships can make that person seem not just accessible, but *sexually* accessible. The availability of these intimate details and images may allow the fan to mistakenly believe that the celebrity is so uninhibited and indiscriminant that she'll have sex with practically anyone, and ultimately this belief seems to work in her favor for a short time. Paris Hilton's fame and name recognition sky rocketed after the unauthorized release of her home video sex tape. In his blog, *Behind the Curtain,* Wayne Parillo writes about his love of Hilton:

> I love Paris because she's in that category of neo-celeb who, if you actually met her, might make you famous. Here's the logic: Paris is famous for being famous—well and for a pornographic video tape—and despite appearances in *[House of Wax]* and an album she is still best known for her, um, how shall I put it . . . she'll sleep with anyone. Now . . . Hilton needs to stay famous for being famous. If you meet her there's a good chance she'll hook up with you. Her making out is the LA version of a handshake. . . . She'll deny it, of course. However thanks to paparazzi and video camera phones SOMEONE is bound to take a picture. Heck someone might even make a video. Insecurity is a powerful thing. She's so insecure that she might prove how powerfully secure she is with herself and how awesome she is in the universe by nailing a stranger.[11]

Because the world has seen the bodies of these celebrities and typically private moments of their sex lives, some fans might feel like the woman would be no more discreet if they were to meet them in person. The seemingly indiscriminate nature of who they expose themselves to lends an air of accessibility to the stars, and the result is that she receives public attention.

A more innocent facet of Internet capability also allows fans to imagine they have a relationship with stars. Celebrities, particularly reality stars, are increasingly authoring their own blogs. This gives them an opportunity to share personal experiences that do not get covered on television. Moreover, it allows fans to comment, and then feel like they are responding to the star in a way that she might actually read and have the opportunity to respond back to. Chaunte LeBlanc of the short-lived MTV reality series *Maui Fever* maintained a blog, in which she discussed her personal life and events that occurred on her show.[12] Perhaps because her personal life and the show were essentially the same thing, everything about LeBlanc's blog smacked of promotion. In one post she directly acknowledges and addresses her fan base: "Well, it seems, after all the drama that unfolded on the last episode, a lot of you have been curious as to why I hooked up w/ Bryce. And to be honest, I did it because he is super cute, and I liked him at the time. And . . . just to throw this out there . . . I am the type of girl that will go anywhere to get what she wants."[13] Later on in the same entry, Leblanc switches topics, but still refers back to comments left to her by her fans: "And there have been more comments on my style. Still flattered . . . As far as shopping goes, I'd say I tend to pick up whatever I think is cute in the boutiques on Maui. One style tip: Get Hula Bronzer, and don't wear too much eyeliner. Although, mascara is always needed. And if you don't feel like wearing makeup, you are beautiful anyway."[14] Here, she is asserting that she notices what her fans tell her and appreciates it. She is also, almost forcefully, maintaining her connection to the circulation of capital and commodification. But her final words are those of acceptance, which seem directly targeted at making the fans feel appreciated rather than alienated. There are twenty-two comments on this entry, and most of them are supportive. Some give advice about Leblanc's relationship issues, some ask questions, and some express admiration. One comment from user "annasarethebest" signifies that Leblanc's tactics of artificial intimacy are working:

> hey chante [sic]. my name is anna and i am from florida. i am in love with the show and you and anna. i swear you and anna, and myself and my best friend masha must be from the same womb. we have been through the same situations you ladies have several times over and yet we always seem to come out looking our best . . . ALWAYS! haha anyways just wanted to let know know i am a huge fan and thinkin of moving to maui to live there with my older brother. hope to see you there![15]

This commenter speaks to LeBlanc in the first person, she shares information about herself, and she insists that her life is parallel to that of the celebrity's. In fact, she is even thinking of moving to the same place (but she's careful to mention her brother as a real connection and reason for the move; she's no stalker).

It is difficult to analyze how real this accessibility to the celebrity is. For the most part, it is probably an illusion, a careful marketing technique. It is not likely that the celebrity blogger will take the time to respond to many comments made by individuals. In fact, she might not even be writing the things that appear on her blog under her name, but there is no way to know for certain. What is real is that someone responds to some of the comments the users leave. In that way, there is a certain amount of access to the star that the fan would not have without the blog because it is possible, in theory, that Chaunte Leblanc does write her own blog, has read the comments, and could respond to one particular person's question. However, fans like "annasarethebest" might imagine there is a stronger connection between themselves and the object of their admiration than there really is.

Twitter, the social network that began in 2006 and has snowballed in popularity since then, has increased this perceived connection between celebrity and fan exponentially. Celebrities frequently discuss twittering during television appearances, perhaps even stopping on the red carpet or at a late night talk show while on camera to tweet to fans about what is taking place at that moment. Those that follow the celebrity on Twitter are then able to receive the tweet and feel as if they are practically there, reinforcing their connection to the star and perhaps also vicariously securing some sort of shared celebrity for themselves. In turn, celebrities often visually and verbally reinforce the perception that they alone send out tweets under their username, unmediated by an assistant or producer, which amplifies the authenticity level of the perceived connection to the audience. Another important facet of Twitter, particularly with regard to the socialite, is that it allows a celebrity to remain in conversation with the public, regardless of whether or not she has a current show, album, or film to promote.

Real and Unreal

Most audiences are aware that some presented aspects of a celebrity are not authentic and that some must be attributes of the real, live human being, but it is often very hard to draw the line that distinguishes the two or decide which way the two are balanced. Michael Newbury discusses Daniel Boorstin's distinction between the "heroes" of the previous century and celebrities of a more modern era by declaring in 1962 that "The celebrity is a person who is well known for his well-knowness."[16] He writes, "For Boorstin, the celebrity reeks of inauthenticity, of cultural declension from a prior age in which 'heroes' were admired for 'greatness in some achievement.'"[17] Also, "Celebrities speak with utmost seriousness to the falsity of modern experience, to the synthetic and mass-produced character of twentieth century consciousness."[18] How to explain that over half a century later, what we know as celebrities have come so much closer to this description than they were at the time these statements were made? And how much more inexplicable is it that with the advent of "reality" television, the stench of inauthenticity is that much more pronounced, as binaries of authentic and inauthentic further collapse?

With reality stars in this early twenty-first century, there is not an overwhelmingly apparent discrepancy between a real life and an on-stage, performed life. Of course there must be something in these stars' private lives that remains private (*mustn't* there?). But with the overabundance of paparazzi snapping photos every time they step outdoors, fans with camera phones able to circulate videos on the Internet in seconds, and tech-savvy BlackBerry® hackers releasing private information like emails, phone numbers, and the contents of rented storage units, it becomes increasingly difficult to tell what was meant to be kept private. And perhaps there is a certain amount of acting in their personal lives, even when the celebrities aren't on camera, because they know information will make its way to the press. So the boundaries of real and staged overlap in an indistinguishable way. Furthermore, these women's careers have been cultivated at such an early age, often before their eighteenth birthdays—for Hilton, since she was fourteen or fifteen—which makes it incredibly difficult for the public to perceive an original personality from which to differentiate the star persona.

In fact, one of the reasons that people are drawn to Paris Hilton and other women like her is because she does *not* make any pretense to great depth of character. They *like* that she is nothing more than what she appears to be and doesn't pretend she is something she is not. For example, *Parterre Box* is a blog centered on, of all things, the opera. One posting includes a YouTube clip of a performance by Anna Netrebko. The blogger, La Cieca, writes, "Note at the end of the clip the lovely and talented Paris Hilton sulking in Richard Lugner's loge."[19] Indeed, in a high culture/mass culture mash-up the camera noticeably focuses on Hilton for a moment. One user writes,

> As far as Paris Hilton is concerned . . . the thing that I LOVE about her is the fact that she never pretends to be anything other than what she is—a trashy, snotty, slutty, blonde who has far too much money at her disposal—and for that I can love her. I would never expect her to understand—or even ATTEMPT to understand art, much less operatic art . . . and so her reaction (or rather, lack of one) at this event does not offend me. She is there as a socialite, not as a supporter . . . for opera.[20]

With some celebrities, superficiality is all that we have to hold on to, and there is the possibility that there really is nothing below the surface, on stage or off. This type of star has an uncanny knack for deflating dimensions, flattening depth until it is paper thin. In *Media Reception Studies*, Janet Steiger provides her take on Richard Dyer's concept of a star's *image* in four categories:

> the star as *persona* (which may or may not be like the "real" person but which is the intertextually constructed notion of the star through a series of films or television programs and which is known, perhaps, only through watching fictional texts); (2) the star as performer (acting ability or how the star plays the roles he or she is assigned); (3) the star as worker/laborer (the professional life of the individual or how she or he negotiates work situations); and, (4) the star in the domestic, *private* sphere (the so-called off-camera life).[21]

The reality starlet complicates these categories by collapsing the text of the celebrity. The star that is famous for being famous has an image that consists of a persona constructed initially without the assistance of film and often *with* the assistance of a television format that claims to be a portrayal of the star's life, which must be somewhat fictional but pretends it is not. Acting ability is not a concern. The concept of the star as worker/laborer is frustrated because the socialite by definition does not seem to work. The concept of reality TV reinforces the perception of the absence of labor by claiming, tongue in cheek, that the camera is merely set up to capture what would have happened anyway. Thus what we see on the screen is not acting, not work, but life being lived.

In the case of Paris Hilton and Nicole Ritchie, the concept of their show *The Simple Life* depends on the portrayal of the pair as incompetent laborers by putting them in situations that would be routine for working and middle class Americans. The women are then shown failing miserably at the tasks they are assigned. An example can be seen from a 2005 episode called "Daycare," when the pair temporarily moves in with a middle class family. The mother of this family drives Hilton and Ritchie to their assigned job, asking them, "Have you ever worked in childcare?" Hilton answers in a heartbeat, "Do we *look* like we've worked in childcare?"[22] Snapping her head back into submission, the woman quickly says no, as if that should have been obvious to her. Despite their inability to work, the show is constructed so that Paris and Nicole generally have enough *social skill* to get out of these ridiculous situations relatively unscathed. Steiger states that "promotion, publicity, films, and criticism and commentary" construct the categories of a star's image.[23] But can one be involved in promotion, publicity, criticism, and commentary without films? They can. They promote and publicize themselves, and others criticize and comment on their personas. Self-promotion can also overlap with the promotion of the socialite's reality show, which portrays her "real" life. But a television show is actually not necessary for this. Paris Hilton just promotes Paris Hilton. In fact, it is believed that as a teenager she hired a publicist to ensure her a spot issue after issue on "Page Six" of the *New York Post*.[24] Because it is difficult to pinpoint the seminal moment of fame for these stars, it is hard to authenticate Hilton's initial entrance into the public sphere. But should this publicity stunt be true, Hilton promoted her own social life and nothing more in order to gain recognition.

It is fitting that reality television programming is the format that has received these celebrities so warmly. The genre shares many qualities with the celebrities themselves, particularly the blurring of reality and fiction. Many viewers of reality television labor under no illusions that they are being presented some sort of true-to-life documentary. The degrees of belief and disbelief fluctuate, but skepticism is always present. Newbury writes, "Celebrities, as Boorstin emphasizes, seem to be conspicuously and essentially enabled and constructed by modern technologies and organizations of cultural production. Where better to look for the false realities embraced by audiences, or, conversely, the capacity of audiences to find their own meanings despite the assertion of dominant, corporate, and commercially motivated interests?"[25] So television, along with the Internet, seem to become interlaced with who these celebrities are, even if it is impossible to precisely pinpoint who they are.

Capital

Some reality TV shows, like MTV's *Maui Fever* and *Laguna Beach*, came under public criticism for portraying high schoolers and college students as beautiful youths with plenty of money and no apparent occupation.[26] The shows center on the act of socializing. If the celebrity socialite thrives on the absence of work, she also paradoxically emphasizes an excess of money. And if the socialite promotes herself, she is often also hawking some product for the public to buy. Paris Hilton, who excels at socializing, is also the stand-out in a group of similar women because she best uses the signification of fame to her advantage. She was, of course, born with an excess of capital into a family that is America's version of aristocracy. But she also excessively owns and promotes the acquisition of *things,* and this has contributed to her success as a celebrity.

One blogger, who wrote in 2007 under the name "chartreuse," analyzes Hilton's connection to commodity in a blog post titled "Why Paris Hilton Is Famous (Or Understanding Value in a Post-Madonna World)." The sections of prose are spaced out with a few high-gloss photos of Hilton, followed by 158 comments by other users, most of them proclaiming the brilliance of the blogger.[27] Chartreuse writes, "I'm amused by Paris Hilton. If Madonna was Marketing 1.0 then Paris Hilton is Marketing 2.0. She's a real life version of what value is and how it is created today. Every web developer should pay attention to her."[28] He goes on to explain how Hilton drops name brands every time she is interviewed despite the fact that she has no contract with the companies, which in turn piques the interest of said companies. Whether or not this is actually the case or whether we *all* just live in a society that identifies by corporate logo is unclear. But this blogger and most of his commenters believe that this is an intentional use of guerilla marketing techniques by Hilton because "[t]he most valuable commodity today is attention."[29] Chartreuse takes the marketing concept even further: "Books, records, movies, etc. don't work for Paris. Because she's actually a platform. Like Digg and YouTube."[30] In saying so, the blogger suggests that her lack of identifiable talent seems not to matter because she is actually integrated into the system of commodification as a walking advertisement.

Many who converse in Internet chat rooms feel that Hilton's fame is due primarily to her good looks and family fortune and seem resentful of that fact. One website called FilmSpot allowed users to submit reviews of films and DVDs online. I was expecting to find reviews for Hilton's films, but instead found reviews of Hilton herself, as if she were the product that was to be rated good or bad. And the reviews were indeed mostly negative. Two examples of such posts, 1) by user Rogue_Beeyatch and 2) CSI_Fiona:

> 1) I have just one question for Paris lovers. Why her? If you people want [a] good actress—you have Charlize Theron. If you want [a] good singer—you have, for example, Shakira. If you wanted a porn star—there is always Jenna Jameson. They are all blond and hot, but unlike Paris they are good in what they are doing and they earned their money by themselves. Paris is one spoiled brat. She is a real proof for a sentence "Some people don't know what to do with all that money." And instead of helping someone, even if it is some fury

> [sic] animal that hot chicks usually [want] to protect, she makes porn videos and stupid parties at home.
>
> 2) She's just one of the many clones on the Hollywood assembly line. They might have different colour hair, eyes but they are all talentless bimbo's who are encouraging anorexia in teenagers. All they can act in is so called reality TV shows like the Simple Life where it looks staged but [then] you think "hang on, no way could Paris Hilton be that good an actress!" The only assets they have are money and looks.[31]

The site also included a pie chart of the percentages of reviews. The largest percentage (27.74 percent) was represented by reviewers that gave Hilton ½ star out of 5 stars. However, the second largest percentage (26.35 percent) was from reviewers who gave her 5 out of 5 stars. Many of the people that gave her 3, 4, or 5 stars had negative things to say about her despite their relatively high ratings. The people that had positive things to say about her often acknowledge that she can't sing or act but don't think that matters. There are, certainly, a handful of users that see Hilton as a successful role model and refuse to be budged by hateful comments posted by people they assume are jealous of Hilton. But many more either dislike her or are somewhat ambiguous about their feelings towards her. So, the majority of Hilton's reviews are negative, yet she has an almost equal percentage of the lowest rating and the highest rating awarded by FilmSpot users. The pie chart is somewhat representative of the sentiments of the American public, which directs at least a portion of its attention to celebrity socialites, whether it is because they have a lot of money or aesthetic beauty, behave outside of societal morality codes, possess or lack talent.

Love to Hate

One common reason people stay tuned to reality shows featuring women like Paris Hilton, Kristin Cavallari, Lauren Conrad, Chaunte LeBlanc, Heidi Montag, Kim Kardashian, or—fill in the name of the socialite of the moment—is that the audience loves to hate these women. In a blog dedicated to "The People We Love to Hate" (with an equally straightforward URL at richfamousandgross.blogspot.com), the blogger, Kelditty, posted an entry about Kristin Cavallari, criticizing her now former relationship with singer/reality star Nick Lachey. Some of the commenters on the entry agreed with the post, but some got into a debate in defense of her makeup choices. Similar to comments about Hilton on other sites, one user wrote these remarks about Cavallari: "HAHAHAHA That is some scary looking makeup. It is ridiculous that this girl is on the cover of Us Weekly. She's famous because her daddy's rich, she lives in a cool place, and she's a bitch . . . amazing. Now I'm not saying I didn't watch Laguna Beach, but why the hell is she still around?"[32] Shortly following this post, another anonymous commenter writes, "I don't see the big deal about her makeup. It's not garrish [sic] or sloppily applied." One user named "meme" takes the time to explain the online situation thusly, "No it's actually not that bad. . . just that some of us dislike her a lot and just like to insult her. You'll find that a lot on this site . . . which makes it fun as [. . .] celeb-bashing. Sometime[s] it gets out of hand."[33] This brief blog entry includes a criticism of a celebrity that people like to hate, a commenter's admission of watching that celebrity's reality show regardless (or because of) one's dislike, as well as an admission that the criticism is blown out of proportion, but that's only because "celeb-bashing" is so much fun.

Fans who find pleasure in "celebrity bashing" also fall under the category of people who "love to hate." When expressed toward a star, this sentiment resembles reveling in the pleasure of loathing something that isn't capable of being harmed by such harsh feelings, a dislike for someone who profits from negative attention so that both the celebrity and the loather win. It is an acceptable form of hatred. This feeling of disgust toward celebrities like the ones I've mentioned commonly crops up in online opinions of reality television cast members. Countless times in my research, I encountered the phrase, "with no discernible talent." But isn't becoming famous without possessing any of the qualities that society deems "talent" a talent in and of itself? The *reality* of the situation is that the ability to succeed against all odds in a capitalist society through marketing an image is actually one of the talents that we, historically, reward most. Another is the ability to socialize, to smooth talk, to sell. Although we may not *like* that we reward those who excel at these things, we do so nonetheless. The hyperreality of the celebrity socialite must have some connection to reality, and in this instance it is to hyperbolically reflect reality back to us. Thus, if we don't like these women without having met them, then our dislike probably has nothing to do with the woman, but rather the systems she represents. The idea of "loving to hate" a celebrity socialite, then, is actually hating what she represents—the privileging of whiteness or class, representations of the female body, the uneven distribution of wealth—but, perhaps masochistically or hypocritically, also being titillated by what she represents.

Andy Warhol, Elizabeth Taylor and Death?

In *Heavenly Bodies*, Richard Dyer writes,

> In part, the fact that the star is not just a screen image but a flesh and blood person is liable to work to express the notion of the individual. A series of shots of a star whose image has changed—say, Elizabeth Taylor—at various points in her career could work to fragment her, to present her as nothing but a serious of disconnected looks; but in practice it works to confirm that beneath all these different looks there is an irreducible core that gives all those looks a unity, namely Elizabeth Taylor. De-

> spite the elaboration of roles, social types, attitudes and values suggested by any one of these looks, one flesh and blood person is embodying them all. We know that Elizabeth Taylor exists apart from all these looks, and this knowledge alone is sufficient to suggest that there is a coherence behind them all.[34]

The celebrity socialite has managed to confound our conception of the star by defying this logic. The proliferation of her two-dimensional image works to undermine the concept that there is any identifiable irreducible core. One signifier that Paris Hilton claims that others cannot is her connection to a more "authentic" star from the golden age of Hollywood, Elizabeth Taylor, who was briefly married to her grandfather. American aristocracy meets Hollywood royalty. Instead of following in her ex-step-grandmother's footsteps, however, Paris Hilton actually more closely resembles another icon, the *portraits* of Elizabeth Taylor painted by Andy Warhol: flat and easily replicated with minor variations, able to be owned in poster form by practically anyone in America.

Fredric Jameson analyzes Warhol and his celebrity portraits in *Postmodernism, or, the Cultural Logic of Late Capitalism*: "The waning of affect is, however, perhaps best initially approached by way of the human figure, and it is obvious that what we have said about the commodification of objects holds as strongly for Warhol's human subjects: stars—like Marilyn Monroe—who are themselves commodified and transformed into their own images."[35] Undoubtedly, much of Jameson's perspective throughout his analysis of *Diamond Dust Shoes* holds true for the celebrity socialite, too. Except that, while the waning of affect with regard to socialites must resonate for many, Paris Hilton and other women like her seem to produce strong emotions in others—disgust, admiration, loathing, adoration—everything *but* indifference because only then will she cease to exist in the public eye. In this way, the hyperreal of the celebrity socialite actually seems to *confirm* and *reinvigorate* the real; she exists primarily because of the reaction against her which reinforces the principles of reality that her audience values, or at least sees as necessary or is nostalgic for. The real, therefore, is not absorbed by the hyperreal in this instance, but stands alongside it made more visible by its presence.[36]

However, it does seem to be true that the popularity of the socialite is waning—at least for now. Whether that is because of a shift in media or public attention, a change in the political atmosphere, the turn from a recession to a depression in which glorification of the wealthy and frivolous elite becomes completely unbearable, or something else entirely remains a question to be taken up elsewhere. In my own observations, the resonance of the reality star seemed to reach an ultimate saturation point upon the death of Anna Nicole Smith in 2007. Every newspaper, magazine, tabloid, and news channel—local and national, entertainment and political, radio and TV—carried stories on her death for weeks. Everyone heard about it, talked about it, had an opinion about it and the market simply could not maintain the same level of interest after her death. That being said, reality television isn't going anywhere soon; it is just too economical to produce. And the celebrity socialite, too, is still alive and well. The *Real Housewives of Orange County*, a Bravo series that began in 2006, follows a generation of women that is surprisingly older than most socialites. They do absolutely nothing that "real" housewives do, yet the series is still running and more successful than ever: the *Real Housewives* is now a franchise with several spinoffs in other cities across the country.

Certainly, Paris et al. used to be in the press much more than they are now. But the figure of Elizabeth Taylor again puts some perspective on the threshold of fame. In an interview for *Cavalier* in 1966, Warhol's states, "I also did movie stars—Marilyn Monroe, Elizabeth Taylor, Troy Donohue—during my 'death' period. Marilyn Monroe died then. I felt that Elizabeth Taylor was going to die, too, after her operation."[37] And Jameson also writes on this concept of death, tying "the deathly black-and-white substratum of the photographic negative" in *Diamond Dust Shoes* to "the emergence of a new kind of flatness or depthlessness, a new kind of superficiality in the most literal sense."[38] Jameson continues, "The great Warhol figures—Marilyn herself or Edie Sedgewick—the notorious cases of burnout and self-destruction of the ending 1960s, and the great dominant experiences of drugs and schizophrenia, would seem to have little enough in common . . . [with that] which dominated the period of high modernism."[39] While I can agree, these statements by both Jameson and Warhol ignore the fact that Elizabeth Taylor *didn't* die young. Although her appearances on screen dramatically diminished, she made important contributions in the 1990s, not the least of which was the attention she brought to the AIDS epidemic. There's something quite different with the way that Dyer constructs her as a star and how Jameson and Warhol read her image. And while Paris Hilton's comparative longevity in this almost always short-lived position of fame is relative to the Warhol paintings rather than the person of Elizabeth Taylor, it is also possible to make a different comparison. Until her death in 2011, Elizabeth Taylor (or more likely someone working for her) still maintained her fan base by sending out information to the public via her Twitter account.

Notes

1. Kristine Styles and Peter Selz, ed., *Theories and Documents of Contemporary Art: A Source Book of Artists' Writings* (California: University of California Press, 1996): 344-45.

2. Although, if one were born famous, it seems as if the notoriety would be attributed to the fame of their parents or the circumstances surrounding the birth.

3. Robert E. Park, "The Natural History of the Newspaper," *The American Journal of Sociology* 29:3 (Nov. 1923): 276.

4. Ibid., 277.

5. Joseph J. Firebaugh, "The Vocabulary of Time Magazine," *American Speech* 15:3 (Oct. 1940): 235.

6. Michael Newbury, "Celebrity Watching," *American Literary History* 12½ (Spring Summer 2000): 272-73.

7. Ibid., 275.

8. Ibid., 275-76.

9. The show *TMZ* is a unique merging of television and Internet. *TMZ* began as a celebrity-watching website, and its success allowed the spin-off into a reality television show in which paparazzi follow celebrities around attempting to get them to talk to the camera. The show highlights the interaction between star and paparazzo, a relationship that is largely obscured in print, the traditional medium of the tabloid.

10. Another example of the female body as public display and promotional tool is the woman who gets increasingly dramatic plastic surgeries and discusses the procedures in detail in the press. These women—Heidi Montag is one example, Tara Reid is another—demonstrate a desire to transform their bodies into a more Hollywood-accepted form of beauty while hoping to retain public interest by staging a "newsworthy" event.

11. Wayne Parillo, "Why I Love Paris Hilton" in *Behind the Curtain*, April 9, 2007, http://wparillo.blogspot.com/2007/01/why-i-love-paris-hilton.html.

12. I would like to briefly acknowledge here the special difficulties that arise when writing on this subject. Odds are, you may be having a hard time recalling both the names of this woman and the show she appeared on. I include Chaunte LeBlanc's story here as not only a testament to the fleeting nature of this kind of fame, but also to indicate how far-reaching these modes of celebrity have become. And it does illustrate how much more one must do to maintain the fame of longer-lasting celebrities. LeBlanc's blog was one of the first of its kind to appear, connecting the TV show to the Internet, connecting fans to the "real" girl on the screen, a practice that is now essentially requisite in the world of reality television. For instance, in the *Real Housewives* series, squabbles over blog contents frequently make their way onscreen. Paris Hilton is clearly an easier case study than LeBlanc is due to her semi-permanence, and it is inevitable that many of the names in this article will be increasingly difficult to remember and new names will continue to arrive.

13. Chaunte LeBlanc, *Chaunte's Blog*, accessed April 9, 2007, http://www.chauntesblog.com.

14. Ibid.

15. annasaresthebest, comment on LeBlanc.

16. Newbury, "Celebrity Watching," 273.

17. Ibid.

18. Ibid.

19. La Cieca, *Parterre Box Presents La Cieca*, accessed April 6, 2007, http://parterre.com/2007/03/opernballistic.html.

20. Il Tenore di Coloratura Superba, comment on La Cieca.

21. Janet Steiger, *Media Reception Studies,* (New York: New York University Press, 2005): 116.

22. *The Simple Life*, "Daycare," TV. Kasey Barrett et. al., March 9, 2005. Los Angeles: 20th Century Fox Television, Bunim-Murray Productions.

23. Steiger, *Media Reception Studies*, 116.

24. Jack Myers, "Who Needs Page Six?" in *Jack Myers Entertainment Report.* Accessed Apr 9, 2007. www.mediavillage.com/JMER_Archive/11-1803ER.pdf.

25. Newbury, "Celebrity Watching," 273.

26. I am sure many of the people on these shows had real jobs at some point, although many teens on *Laguna Beach* did seem supported by their parents. The criticism is that the show claims to follow the real lives of these people and yet they never work. Editing is of course the primary reason that reality television cannot portray reality.

27. While researching, I posted websites to my del.icio.us account, a public website that allows users to bookmark their favorite sites so they come back to them later. This space also allows other users to see what you've bookmarked so that they can save the sites for themselves. I posted 25 websites about the celebrities to del.icio.us. Two of them were also saved by three other people, and most were not saved by anyone else. This particular blog was saved by 244 other users between the dates of 3/28/07 and 4/6/07, which is some indication of the interest in Paris Hilton and marketing.

28. chartreuse,"Why Paris Hilton Is Famous (Or Understanding Value in a Post Madonna World)" in *chartreuse: internet media network management*, accessed April, 2007. http://chartreuse.wordpress.com/2006/09/18/why-paris-hilton-is-famous-or-understanding-value-in-a-post-madonna-world/.

29. Ibid.

30. Ibid.

31. Filmspot.com, "Review Paris Hilton," accessed April 6, 2007, www.filmspot.com/people/207608/paris-hilton/reviews.html.

32. Kelditty, "MTV Reality Meets for Drinks" in *The People We Love to Hate: Bitch Slapping Celebrities One Hit at a Time*, accessed July 2, 2010, http://richfamousandgross.blogspot.com/2006_03_12_archive.html

33. Ibid., my ellipses in brackets.

34. Richard Dyer, *Heavenly Bodies: Film Stars and Society* (New York: Macmillan Education Ltd, 2004): 9.

35. Fredric Jameson, *Postmodernism, or, The Cultural Logic of Late Capitalism* (North Carolina: Duke University Press, 2003): 11.

36. While scholarship often addresses the politics of the 1990s and follows reality television to its height during the George W. Bush presidency, an interesting moment occurred during the McCain/Obama election: in a campaign ad, McCain linked Obama to Paris Hilton and Britney Spears by juxtaposing clips of the two women with cuts of Obama in front of animated crowds of people, implying the future president's popularity was due to a cult of celebrity devoid of political substance. Paris Hilton was given the opportunity to rebut the campaign ad in a surreal, comical spoof on funnyordie.com. She began by saying, "Hey America, I'm Paris Hilton and I'm a celebrity, too. Only I'm not from the olden days [like McCain] and I'm not promising change like that other guy [Obama]. I'm just hot. But then that wrinkly, white haired guy used me in his campaign ad, which I guess means I'm running for President!" However, Hilton then goes on to describe her energy policy, which entails "limited off shore drilling with strict environmental oversight, while creating tax incentives to get Detroit making hybrid and electric cars. That way the off-shore drilling carries us until the new technologies kick in which will then create new jobs and energy independence. Energy crisis solved! I'll see you at the debates, bitches."

37. Kennneth Goldsmith, Reva Wolf, and Wayne Koestenbaum, *I'll Be Your Mirror: The Selected Andy Warhol Interviews 1962-1987* (New York: Carroll and Graf Publishers, 2004): 99.

38. Jameson, *Postmodernism,* 9.

39. Ibid., 14.

Chapter 2
Simulating Supper: Serving Up TV Dinner

Glenda Shaw-Garlock

Introduction

This essay explores the juxtaposition of performance, food, and family values using the concept of "simulation and simulacra," formalized by French social theorist Jean Baudrillard as well as the concept of "pseudo-events" conceived by historian Daniel Boorstin in addition to Donald Horton and Richard Wohl's theory of "para-social" relationships. Applying these theories to contemporary *food television* we may trace the gradual shift toward food consumption as simulation and pseudo-event and para-social interaction which began with such early television personalities as Julia Child and Graham Kerr and continued through the genre defining format of Emeril Lagasse, and persists within the increasingly sensuous food and program styling of Rachael Ray, Nigella Lawson and Giada De Laurentiis.

The Rise of Food Programs, Food Television & Celebrity Chefs

Food-based television programming has been around almost as long as broadcasting itself beginning with X. Marcel Boulestin broadcasting on the BBS in 1937.[1] James Beard first hosted a food show in the Unites States in 1946 on a show called "I Love to Eat."[2] Dione Lucas was the first to introduce the American viewing public to the French culinary arts in the 1950s, but it was Julia Child who redefined food television with her classic food program, "The French

Chef' in the early 1960s.[3] Graham Kerr's was the first truly international culinary food, gaining stardom first in New Zealand and Australia in the early 1960s and then eventually in North America through the very successful "Galloping Gourmet," made in Canada by the CBC, from 1969-1971.[4] Martha Stewart took the spotlight in the 1980s and interest in food and lifestyle programming continued to grow throughout the 1990s, due in large part to the emergence and ongoing success of Food Network.

The Food Network (first known as The TV Food Network or TVFN) began broadcasting on November 23, 1993.[5] In 1999, the network boasted forty million viewers and in 2010 it now reaches more than ninety-eight million North American households. Headquartered in New York, The Food Network continues to expand internationally exporting its brand to more than 150 countries, which includes twenty-four hour networks in Great Britain, India, Asia and Africa.[6] Food Television is also developing a new channel, The Cooking Channel, which launched May 31, 2010.[7]

Today, the effervescent Rachael Ray exclaims, "Yum-O!"; "that's money!" proclaims Guy Fieri; "allez cuisine!" commands the chairman of Iron Chef; and finally, who is not familiar with Emeril Lagasse's trademark "Bam!"? All of these celebrity chefs are associated with Food Network's most popular programs and are far removed from the humble public television beginnings of Julia Child and Graham Kerr. Food television moved from daytime to primetime in the early 1990s and since then foodie television has gone mainstream: Bravo carries "Top Chef," TLC runs "Cake Boss," Fox has "Hell's Kitchen," while ABC has picked up "Jamie Oliver's Food Revolution."

Because Emeril Lagasse is Food Network's most widely recognized and enduring celebrity chef, his trademark program format provides a compelling site of inquiry and analysis for this work. Lagasse is a cultural phenomenon and he positioned himself early as a rising Food Network super star. The "Essence of Emeril" began taping in 1994 and was The Food Network's top rated show for many years. "Emeril Live" commenced production and replaced "Essence of Emeril" in January, 1997. In April, 2010, "The Emeril Lagasse Show" began airing on ION Television.[8]

Lagasse's personality is gregarious, down to earth, and over the top, as he exclaims, "this [cooking] ain't rocket science!" When you watch an episode of "Emeril Live," "Essence of Emeril," or "Cooking with Emeril," the first thing you note is a program format that combines characteristics of theatre performance and a dinner party, resulting in a complex form of dinner theatre. Lagasse also features a live band and studio audience, while he entertains, philosophizes, teaches, and flirts (with the audience and viewer) as he prepares marvelous three to six course meals that the live studio audience is invited to eat. In short, he turns the everyday family meal into and an exuberant raucous media event. Frederick Kaufman notes that as people cook less and less, they tend to ogle food shows more and more.[9] Adema suggests, "Food television is not about eating: It is about watching food and being entertained by the personality."[10] In short, watching dinner has been become a substitute for and perhaps even more pleasurable than actually cooking and eating dinner. The increasing popularity of watching simulated TV dinner represents an instance in which "the fictional takes precedence over the actual"[11] and in Baudrillard's words perhaps the family dinner "is no longer in the order of appearance at all, but of simulation."[12] In the following section Baudrillard's conception of simulation and simulacra is considered in more detail and each step in the process of simulation is considered in relation to various aspects of food television programs and personalities.

Simulation, Simulacra; Pseudo-Events; and Para-Social Relationships

Conceptualized by French social theorist Jean Baudrillard, simulation is accomplished through the following specific steps:

1. Simulation begins by reflecting a basic reality;
2. Simulation continues by hiding and subverting this reality;
3. It then proceeds to conceal the absence of the foundational reality;
4. Until finally, it bears no resemblance what-so-ever to the basic reality: "it is its own pure simulacrum."[13]

Simulacra and simulation describes how an individual's everyday lived experience can be rendered non-real, or hyperreal, through the creation of a false environment which acts upon the individual in such a way as to make simulation seem like reality. Baudrillard cites Disneyland as an ideal example of simulation, " . . . a digest of the American way of life, panegyric to American values, idealized transposition of a contradictory reality."[14] He further suggests that the postmodern experience is one in which signs (anything which carries or conveys meaning) are broken down and disconnected from reality. Understood this way, simulation describes a situation wherein signs no longer represent reality.

The notion of signs dis/representing reality resonates with Daniel Boorstin's concept of pseudo-event or "synthetic novelty"[15] which describes an event which somewhat (but not wholly) deceives the viewer and involves four characteristics. First, the pseudo-event never happens by chance; rather, it is carefully formulated, designed, and executed. Second, the pseudo-event is typically carried out with media and reproduction in mind and the relative success of an event is evaluated by the degree to which media show interest and disseminates the message. The third characteristic of the pseudo-event is particularly resonant with Baudrillard's concept of simulation and simulacra in that the relationship between the underlying reality and the pseudo-event is ambiguous: "Without some of this ambiguity a pseudo-event cannot be very interesting."[16] Finally, pseudo-events are intended to reinforce or "be a self-fulfilling prophecy."[17] As reality is rendered unstable and ambiguous and eventually gives way to complete simulation, "a play of illusions,"[18] we may say that the viewing audience member and food television celebrity chef are engaged in a para-social relationship. Horton and Wohl describe this relationship as follows,

> One of the striking characteristics of the new mass media—radio, television, and the movies—is that they give the illusion of face-to-face relationship with the performer. The conditions of response to the performer are analogous to those in a primary group. The most remote and illustrious men are met as if they were in the circle of one's peers; the same is true of a character in a story who comes to life in these media in an especially vivid and arresting way. We propose to call this seeming face-to-face relationship between spectator and performer a para-social relationship.[19]

Thus, the illusion of an intimate face-to-face relationship between spectator and performer (also called a "persona") is constructed. It is illusory because the relationship is inevitably "one-sided, nondialectical, [and] controlled by the performer."[20]

Simulation Begins by Reflecting a Basic Reality

Applying the principles of simulation/simulacra and the characteristics of the pseudo-event and para-social interaction to popular food television suggests that analysis must begin with identification of the *basic reality* that the television program attempts to subvert.

Unlike visiting the physical location of Disneyland, the television viewer of cooking shows such as "Emeril Live" seem to be invited to construct impressions associated with meal preparation and consumption through a pseudo-event being acted out in the para-social relationship between the television spectator and the chef on the television screen, thereby transforming dinner into a fabricated performance. I suggest that the foundational reality a televised cooking show engages with is that of the traditional family dinner associated with a social order that appeals to a nostalgic understanding of the past. Pauline Adema connects the appeal of food television to "hunger for comfort and security, traditionally sated in the home kitchen and encoded in home-cooked food."[21] Lise Shapiro Sanders associates the popularity of such food celebrities as Nigella Lawson, in part, to a nostalgic longing for domestic femininity.[22] Elizabeth Nathanson suggests that contemporary domesticity is "ambivalent, positioned between the forward-moving hands of the market and the backward-looking glance of nostalgic desire for a fantasy past in which women had all the time in the world."[23] Thus, the food preparation site, located on the set of food television programs, evokes utopian memories of mealtime with a traditional mother-figure nestled reassuringly inside the kitchen. Where once the family meal was a common and daily event performed by mother or wife, today it is "earning 'rarity values,' a nostalgic value ascribed to something based on its scarcity."[24] This shift of value may partly account for the Food Network's success in that as society spends less and less time engaged with actual meal preparation it grows increasingly nostalgic and hungry for the traditions (and kitchens) of the past. The regularized appearance of such chefs as Emeril Lagasse may be likened to the ritualized events of para-social relationships that come to be "to be counted on, planned for, and integrated into the routines of daily life."[25] Thus, the regular interaction between viewer and celebrity chef begins to mimic the rhythms and routines of the family dinner of the past. Horton and Wohl suggest that audiences may regard the television persona as a friend, counselor, comforter and model and I suggest that in the case of the television chef we may also add: symbolic maternal figure.

In summary, the basic reality subverted in the process of simulation is traditional middle class values, which include families sitting down together eating a wholesome meal lovingly prepared by mother. This reality is fabricated into a pseudo-event that is performed by a food television chef, studio audience, and television viewer in form of a contrived para-social interaction.

Simulation Continues by Hiding and Subverting Foundational Reality

Baudrillard states, "When the real is no longer what it used to be, nostalgia assumes its full meaning."[26] The second phase of simulacra and simulation concerns the way in which a basic reality (e.g., traditional family values) is subverted and hidden. An examination of the genre of cooking shows and its historical icons may help determine how this process is accomplished. The history of food television begins with classically trained Graham Kerr and Julia Child, who epitomized high-culture *haute* and French cuisine during the 1960s and 1970s. Indeed, the television chef has been in the public consciousness from the very earliest broadcasting years.

Julia Child burst onto the scene in the late 1960s with her very popular, "The French Chef,"[27] overlapping somewhat with Graham Kerr's "The Galloping Gourmet." Child's program is generally regarded as the official beginning of food television programs.[28] Kerr and Child prepared complex dishes they learned while studying at elite cooking schools. Their approach to food preparation was deferential and elitist. I do not mean elitist in a derogatory sense, rather in the sense that the language of cooking, its gourmet ingredients, flavors, and culinary techniques reinforces a distinction between the social classes. Child and Kerr, while enormously popular, did not invite the same level of public identification with the working class as contemporary television chefs seem to evoke. Further, there was no attempt to reconfigure a reality located somewhere outside of the program set. The audience understood that Child and Kerr were representing something other than social (and culinary) values of middle class North America.

In contrast, Emeril Lagasse signifies a chef very different from the elite gourmets of the 1970s. "Emeril literally and linguistically deconstructs what have traditionally been elite foodways."[29] He is understood to be an everyday man, a blue-collar guy with a "short order cook moxie" whose cooking style has been characterized as "loose to the point of sloppy" by some critics.[30] Lagasse utilizes the main strategies of intimacy building for creating para-social relationships outlined by Horton and Wohl: replication of the gestures, conversational style and an inviting atmosphere associated with face-to-face interaction.

Lagasse became famous for cooking creative Creole-Acadian (he is of French-Canadian and Portuguese heritage) and dishes that reflect the bayou's culinary tradition while adding on a contemporary flair. Using uncomplicated ingredients, often available at local supermarkets, he employs an indelicate approach to cooking. Rather than precisely seasoning a dish, Lagasse literally throws his spices while shouting, "Bam!" or "Kickin' it up a notch!" The audience responds by shouting their appreciation and delight and collectively gasps and sighs as each dish is dramatically garnished and is passed out for them to sample. Lagasse has also developed many acclaimed restaurants close by or within highly fantastical locations. Many of the locations share many of the same features as Disneyland's simulated environment. For example, Emeril's restaurant *New Orleans Fish House* is located at the MGM Grand Hotel in Las Vegas. He also owns a steak house in Las Vegas and still one other in Orlando, Florida at Universal Studios.[31] I suggest that these fantastical locations further situate Lagasse within a discourse of simulation and simulacra.

As a popular culture icon, Lagasse is attached to complex imagery. Lagasse's image is a simplified representation of a person (comforting paternal/maternal figure), a place (the family kitchen), referencing a historical period (nostalgic past). The promotion of this pared down representation using simulation, pseudo-event techniques (set, script, performer, sound and lighting and so on), and para-social relationship strategies (gesture, conversational style, atmosphere creation) are the means by which "family values" and "nostalgic past" become subverted and hidden.

Simulation Proceeds to Conceal the Absence of the Foundational Reality

Juxtaposing traditional family values of the past with highly contrived and stylized performance techniques is how the basic reality, in the third phase of simulation, is rendered unrecognizable and absent from the foundational reality from which it began.

"Emeril Live" opens with a hip studio band playing music that reflects the ethnicity of the culture associated with the food about to be prepared. This space signifies the global kitchen perspective embodied in his shows. Indeed, Lagasse typically begins each show by teaching a cultural lesson related to the ethnic food that he is about to prepare. We may say that Lagasse stands in for the maternal figure in the contemporary kitchen. The audience, his pseudo-family, is invited into the kitchen to learn about cooking and global culture. This contemporary cooking and looking ritual replaces the tradition of handing down recipes from mother and/or grandmother to daughter. For Horton and Wohl, the apparently intimate face-to-face association between performer and viewer is a distinguishing feature of the mass media generally, but it is especially powerful and recognizable in the "personality" program of which food programming and celebrity chefs are part. Lagasse seeks to connect intimately with his viewership in the way described by Horton and Wohl, as he "often faces the spectator, uses the mode of direct address, [and] talks as if he were conversing personally and privately."[32] Adema describes the compelling nature of Lagasse's program this way, "Here is a man who is cooking for you, looking and speaking to you, and he is passionate and entertaining: This is many women's dream."[33]

Emeril creates a pseudo kitchen; a space commonly associated with the food sharing, family, and formation of childhood food memories.[34] Indeed, families meeting together to dine is a practice which dates back two million years.[35] However, Lagasse's pseudo-family does not sit deferentially and observe and learn; rather, chef and audience interact, play, shout and vocalize to each other throughout the performance. Moving artfully and quickly from dish to dish, Lagasse stirs up the audience until they resemble fans at a rock and roll concert. Adema suggests that Lagasse's kitchen may represent a reworking of gender roles, wherein cooking dinner is no longer regarded as the exclusive domain of mother or woman, "transforming cooking from an essentially feminine chore to a glamorous hobby for men . . . "[36] In short, the preparation of food transforms into an elaborate theatre performance with actors playing particular roles (persona, studio audience, home spectator, etc.). At this point, simulation has succeeded in replacing the real event. The physical production and consumption of food has been replaced with a dynamic gender bending pseudo-event that no longer resembles the traditional family gathering at all. It is at the point where the substitution of reality or performance that simulation completely wipes out the foundational reality.

Simulation Bears No Resemblance Whatsoever to the Basic Reality

In its final stage, simulation bears no likeness to the basic reality from which it began. Hyper-reality is described as a condition whereby "origins are forgotten, referents lost, and simulations begin to precede and engender reality."[37]

We have seen that food television began as an instructional format meant to introduce audiences to the high brow culinary techniques associated with French cuisine. As the popularity of Food Network began to rise, food programming seemed to connect with family values and the intimacy of home cooking through the personality program and the intimacy created by celebrity chef. However, twenty years after Lagasse burst onto the food television scene, contemporary food television increasingly bears no resemblance whatsoever to the instructional and nostalgic programming which preceded it as it.

Although contentious, food pornography is a term which seems to be increasingly associated with food television and food culture more generally. For Richard Magee, food pornography describes a focus "on food or sex as a performance, and like all performance, are designed as a voyeuristic exercise."[38] Food takes on a fantasy dimension that, in the words of Molly O'Neill, is "so removed from real life that they cannot be used except as vicarious experience."[39] Frederick Kaufman argues that both food-porn and sex-porn attend to the most basic of human desires and both tend to romanticize and debase them simultaneously.[40] As is the case in sex pornography, the sumptuous food offerings of gastro-porn appear effortless, yet is impossible to replicate at home, and "unlike home cooking, TV cooking builds to an unending succession of physical ecstasies, never a pile of dirty dishes."[41] As is the case in illusory para-social interaction, the role enacted by the persona, or celebrity chef, is an " . . . idealized version of an everyday performance . . . [a] para-social approximation of an ideal pattern, not often, perhaps never, achieved in real life."[42] Bob Tuschman, Food network's senior vice-president for programming and production, says the following about casting the best food personalities,

> Even when we're doing food television it still has to be great television . . . And it is dependent on having great stars, the person who walks into the room and you cannot take your eyes off them. You are enthralled. When I met Rachael Ray, I had the same feeling. When I met Giada De Laurentiis, I felt the same thing.[43]

Indeed, Food Network stars Gaida De Lauretiis, Nigella Lawson, and Rachael Ray are popular as much for their accomplishments as chefs as for their provocative appearances and sensual styling, attributes that do indeed make it difficult for the audience to take their eyes off of them. From erotically scripted expressions (Nigella Lawson describes one meal as a "chin-dripping, love-soaked primal feast"[44]) to the enthralling closing shot of the chefs passionately eating their finished foods, sexuality seems to pervade these cooking shows. Costumed in body conscious clothing, the camera tends to linger erotically upon their faces, curves, fingers, and mouths—oft times not so subtly mimicking the production values of sex porn.[45]

Rebecca Swenson[46] argues that the Food Network represents a "channel of discourse" and a site which reflects dominant attitudes about gender in/and the kitchen. In some analyses of the highly eroticized chefs such as Nigella Lawson, her programs are read as complex texts that seem to reject the patriarchal oppression of kitchen even as they cleave to the pleasures of domestic comfort (i.e., food, cooking, eating) much as we may pursue the pleasures of sex while rejecting debasement associated with sex pornography.[47] Taken together we see that when food television is read as food pornography and/or as a provocative challenge to patriarchal traditions, these new programs subvert food television as culinary instruction as well as food television as nostalgic gesturing toward some idealized past.

Contemporary food television reflects a complex relationship between performance, food and family values. Baudrillard's concepts of simulation and simulacra assist our understanding of the peculiar phenomena whereby the television viewer, studio audience and television host engage in highly contrived pseudo-events that are in fact, instances of hyper reality. Carefully mapping each step in process of simulation onto contemporary food television programs we see how the physical and material reality of food preparation and consumption has gradually become subverted into a synthetic event that transforms eating into a voyeuristic (and some argue pornographic) form of visual entertainment.

Notes

1. Stephen Mennell *All Manner of Food: Eating and Taste in England and France from the Middle Ages to the Present* (Oxford: Blackwell, 1995), 346.

2. Frances Bonner, "Early Multi Platforming," *Media History* 15, no. 3 (2009): 347.

. 3. Pauline Adema, "Vicarious Consumption: Food, Television and the Ambiguity of Modernity," *Journal of American & Comparative Cultures* 23, no. 3 (2000): 114.

4. Bonner, "Platforming," 348

5. Barbara Shecter, "Emeril Lagasse the Marketable Cook," *National Post*, June 21 1999, C.04; Adema, "Vicarious Consumption," 113-24

6. Anonymous, "Harrah's Entertainment Announces 2010 Food Nework Atlantic City Food and Wind Festival," *PR Newswire*, April 6 2010.

7. Michael Hill, "Food Fans to Get Edgy New Channel," *The Oklahoman*, April 28 2010, 4D.

8. "Chef Emeril Lagasse and ION Television Break Bread With New Original Series." Emeril.com, accessed July 1, 2010. http://www.emerils.com/newsroom/448/chef-emeril-lagasse-and-ion-television-break-bread-with-new-original-series/.

9. Frederick Kaufman, "Debbie Does Salad: The Food Network at the Frontiers of Pornography," *Harper's Magazine*, October 1 2005, 56.

10. Adema, "Vicarious Consumption," 116.

11. Donald Horton and Richard R. Wohl, "Mass Communication and Para Social Interaction: Observations on Intimacy at a Distance," *Psychiatry* 19 (1956): 215.

12. Jean Baudrillard, "Simulacra and Simulation," in *Selected Writings*, ed. Mark Poster (Stanford: Standford University Press, 1988), 170.

13. Baudrillard, "Simulacra and Simulation," 170.

14. Baudrillard, "Simulacra and Simulation," 172.

15. Daniel J. Boorstin, *The Image or What Happened to the American Dream* (New York: Atheneum, 1962), 21.

16. Boorstin, *The Image*, 23.

17. Boorstin, *The Image*, 23.

18. Baudrillard, "Simulacra and Simulation," 171.

19. Horton and Wohl, "Mass Communication," 215.
20. Horton and Wohl, "Mass Communication," 215.
21. Adema, "Vicarious Consumption," 113.
22. Lise Shapiro Sanders, "Consuming Nigella," in *Feminism, Domesticity, and Popular Culture*, ed. Joanne Hollows Stacy Gillis, (New York: Routledge, 2009), 151.
23. Elizabeth Nathanson, "As Easy as Pie: Cooking Shows, Domestic Efficiency, and Postfeminist Temporality," *Television New Media* 10 (2009): 328.
24. Adema, "Vicarious Consumption," 118.
25. Horton and Wohl, "Mass Communication," 216.
26. Baudrillard, "Simulacra and Simulation," 171.
27. Brenda Wineapple, "Crème De La Crème," *Women's Review of Books* 15, no. 3 (1997): 10-12.
28. Bonner, "Platforming," 349.
29. Adema, "Vicarious Consumption," 117.
30. Anonymous, "Emeril," *People Magazine* 50, no. 24 (1999): 102.
31. Rana Dogar and Damon Darin, "Recipes That Grandma Never Made," *Forbes Magazine*, 1996, 190-195.
32. Horton and Wohl, "Mass Communication," 215.
33. Adema, "Vicarious Consumption," 116.
34. Gill Valentine, "Eating In: Home, Consumption, and Identity," *Sociology Review* 47, no. 3 (1999): 491-525.
35. Margaret Visser, *The Rituals of Dinner: The Origins, Evolution, Eccentricities, and Meaning of Table Manners* (Toronto: Harper Collins, 1991).
36. Adema, "Vicarious Consumption," 116.
37. Douglas Kellner, *Baudrillard: A Critical Reader* (Oxford: Blackwell, 1994), 194.
38. Richard M. Magee, "Food Puritanism and Food Pornography: The Gourmet Semiotics of Martha and Nigella," *The Journal of American Popular Culture* 6, no. 2 (2007). Accessed June 15, 2010.
http://www.americanpopularculture.com/journal/articles/fall_2007/magee.htm
39. Molly O'Neill, "Food Porn," *Columbia Journalism Review* September-October (2003): 39.
40. Kaufman, "Debbie Does Salad," 56.
41. Kaufman, "Debbie Does Salad," 57.
42. Horton and Wohl, "Mass Communication," 222
43. Quoted in Kaufman, "Debbie Does Salad," 60.
44. Magee, "Food Puritanism and Food Pornography."
45. Kaufman, "Debbie Does Salad"; Richard M. Magee, "Food Puritanism and Food Pornography."
46. Rebecca Swenson, "Domestic Divo? Televised Treatments of Masculinity, Femininity and Food," *Critical Studies in Media Communication* 26, no. 1 (2009): 36-53.
47. Charlotte Brunsdon, "Feminism, Postfeminism, Martha, Martha, and Nigella," *Cinema Journal* 44, no. 2 (2005): 110-16; Magee, "Food Puritanism and Food Pornography." *The Journal of American Popular Culture* 6, no. 2 (2007). Accessed June 15, 2010; Sanders, "Consuming Nigella," 151-163.

Chapter 3
Imitation meets Simulation: *Seinfeld*'s George Costanza as The "Original" Larry David

Naomi L. Fosher

Today's postmodern society is immersed in television, video games, computers and the internet—all virtual platforms which mediate the nature of reality and show their audiences the world they create. These worlds are copies of reality, but are they really? In this hyperreal world there is a procession of precessions of simulations which make it impossible to prove the real. There are spaces in which *a real* does try to prove itself as *the real*, but we find that the real is always already a simulation.

Reality television is one cultural phenomenon which tries to prove itself as real and much scholarship has been written on the nature of its "reality." But what about more subtle genres which preach verisimilitude, such as television sitcoms? The complex layers and interplay of the hyperreal are worth exploring further in these two American television shows: in the cultural phenomenon *Seinfeld* and *Seinfeld* co-writer Larry David's own show *Curb Your Enthusiasm*. After nine seasons of *Seinfeld*'s popularity, Larry David in his own show *Curb Your Enthusiasm* stars as himself, even though George in *Seinfeld* was already a character based on David himself.[1] David, as a simulation of a simulation, ends up being a caricature of himself because of the success of the George Constanza character in *Seinfeld*. The similitude and relation of both shows' characters and their proclaimed proximity to the real problematizes the existence of a real as *real*. I will explore the complex layers of simulacra and interplay of the hyperreal through these two American television shows and work to reveal, by using theorist Jean Baudrillard's work, that the category of the "real" has now been drained of its ontological meaning.

First, a brief look at the condition of the real and an exploration into Baudrillard's theory of simulacra and hyperreality is necessary. Next the phenomenon of how the character of Constanza seems to end up becoming the original and David the copy is analyzed. I then explore the unscripted "reality" of *Curb Your Enthusiasm,* what has been called "comic verité" or "mockumentary" style. I look at the complex layers of the seventh season of *Curb Your Enthusiasm* when the *Seinfeld* "reunion show" (which is really a simulation of a reunion show) is embedded in the *Curb Your Enthusiasm* season finale. I end with viewers' homages to the real as they fixate on trying to find the "real" though there is no "outside" to the show anymore. This fetishization of the real, evident in essays, interviews and books, further documents the loss of the real in the complex intertwining meta-layers of simulations.

A contemporary view of reality explored in great depth by French theorist Jean Baudrillard problematizes the condition of the real. For thinkers in the past who explored the idea of reality, like Plato for instance, the real still existed. Plato, in "Book X" of *The Republic*, is skeptical of mimesis because each successive copy or representation leads farther from the original, but he still finds that the original can be traced back by the linked path of the successive copies.[2] Baudrillard recognizes a difference in the current situation of our postmodern world. In his book, *The Vital Illusion*, Baudrillard discusses the circumstances of the postmodern real. In the section, "The Murder of the Real," he explains that the real ceases to exist: "The corps(e) of the Real—if there is any—has not been recovered, is nowhere to be found. And this is because the Real is not just dead (as God is), it has purely and simply disappeared."[3] Baudrillard continues to explain this current state by quoting novelist Elias Canetti, who writes, "As of a certain point, history was no longer *real*. Without noticing it, all mankind suddenly left reality: everything happening since then was supposedly not true; but we supposedly didn't notice. Our task would now be to find that point, and as long as we didn't have it, we would be forced to abide in our present destruction."[4] This statement reveals that over time, copies replicate other copies to a point where the original is not traceable. The history of the copies is not real because there are only copies of copies without a material original. This is the nature of simulacra: it is not a representation, but a duplication which replaces the real and thus exists as hyperreal. Also, as Barry Brumett points out, the simulation replaces the need for a real, "Simulation is an experience in which the copy is just as good as reality, even preferable because it is reproducible and thus repeatable."[5]

Baudrillard also ascertains that neither duplication nor artificiality is the issue, but instead the complete replacement of what was real:

> It is no longer a question of imitation, nor duplication, nor even parody. It is a question of substituting the signs of the real for the real, that is to say of an operation of deterring every real process via its operational double, a programmatic, metastable, perfectly descriptive machine that offers all the signs of the real and short-circuits all its vicissitudes.[6]

The issue of the simulacra in our hyperreal order is that it is untraceable and in searching for a real, we instead substitute the signs of the real *for* the real. The real which we believe is real, and the continual repetition of this condition, is a simulation which has now "short-circuited" and destroyed any former real.

The interplay of *Seinfeld* and *Curb Your Enthusiasm* complicates the perception of a real because the shows offer a simulation of a simulation which plays with the idea of hyperreality. The shows do not just consist of simulations which lead back to a "traceable real." Instead, played out through both George Costanza and Larry David, the shows present us with a copy or approximation of simulacra rather than giving us the "real" thing in a world of copies.

George and Larry Meet, For Real

While David is technically the real George Costanza, it is George's position as a copy of David that works to make David a simulation and George the original or real. With *Seinfeld* being such a cultural phenomenon, people watched George and all his mannerisms for nine seasons of the show. The mannerisms, personality, situations and sayings of George become reinscribed as part of George himself. Even though the mannerisms, personality, situations and sayings were originally all parts of David, they have been simulated and then become assimilated with the simulation. In their essay, David Lavery and Marc Leverette point out an example of both George and Larry's neuroses about urinating: George is a "stall man," scared of urinals ("The Note," 0301—notably written by David); Larry injures his back while trying to urinate sitting down in the middle of the night in "The Weatherman" (0404).[7] Both characters are also fanatical about their clothing. Again, in the "The Weatherman" episode, Larry refuses to wear an adhesive name tag at the "Leo Funkhouser Testimonial" as it will "leave little paper remnants and ruin the suede" of his jacket.[8] That neurotic response is reminiscent of George's neurotic response to his broken zipper in his first session of therapy in the *Seinfeld* episode "The Pick" (0413).[9] He becomes fixated on the zipper, can't respond to any questions in the session and his therapist has to try to help him fix it. These are just a few, quick examples of the many mannerisms shared between George and Larry and it is evident that George as a simulation is not a simulation any longer—through the process of viewing George on television he becomes the real David.

The layering of the hyperreal becomes more complex as *Curb Your Enthusiasm* works to inform and complicate the former relationship of George and

David. The real David has been replaced by the simulation George, but then the character of Larry on *Curb Your Enthusiasm* is also a simulation of David. In essence there is a collapse and, as Baudrillard would say, implosion of simulations as the character of Larry simulates David who is a simulation of George. This is the messy hyperreality Baudrillard is discussing where there is a layering of simulations and they become hyperreal, and therefore refer to no original. And in this layering of simulation, each of the doublings works to render any potential real as artificial. Baudrillard states:

> Everywhere we live in a universe strangely similar to the original—things are doubled by their own scenario. But this doubling does not signify, as it did traditionally, the imminence of their death—they are already purged of their death, and better than when they were alive; more cheerful, more authentic, in the light of their model, like the faces in funeral homes.[10]

By doubling something, the original and the copy are not dead, but first artificial and then hyperreal. This is so because in their artificially, as they are beyond the death of the real, they are *hyper*real or beyond real. It is in this way that we see how, though the process of simulation, all the copies change the real to simulations. The artificial nature of the simulations is also what creates caricatures.[11] In *Curb Your Enthusiasm*, the "mockumentary" or mock-documentary/improv style of the show reinforces Larry and David as caricatures because the unscripted nature is a studied formlessness which has an opposite effect and becomes too scripted, stilted and formed. David, essentially, becomes a caricature of himself through his role "as himself" in *Curb Your Enthusiasm.*

This idea of starring as oneself complicates the idea of *self.* If you really starred as yourself, you would be your real self, or one would think anyway. But in playing yourself, you become something else. David explains this complication on *60 Minutes* when he discusses his character on *Curb Your Enthusiasm*: "I love the character. At first I didn't realize it was gonna be a character. I just thought I was gonna be doing me. And eventually as I kept writing it, something emerged that was not quite me but a version of me."[12] David originally was just going to play himself, but in doing so he ended up becoming some part of himself on screen. Since the simulation of George already existed previous to the start of *Curb Your Enthusiasm* in 1999, David was already in a context to be a caricature of himself on screen. Since he plays himself on his show, and there are more copies of copies, this strengthens the caricature situation further so that both the "real" David and the character Larry are caricatures of the George character.

Beyond Larry's and David's state as caricatures, David would not exist without the simulations. A prime example of this argument is that of Marilyn Monroe and Norma Jean: Marilyn Monroe created Norma Jean. Norma Jean is created after the existence of Marilyn Monroe, after the cultural fetishization. Without Marilyn and the Marilyn lens, we would not have even thought about

Norma Jean. This is also true about Larry David the man: without George Costanza and then later the Larry David character, he would not exist. David needs the lens of George and of Larry to even exist; his neurotic self who always takes one step too far, who is now fetishized, is dependent on the existence of the others.

The stereotypical "Larry David" (whatever that means) always takes everything one step farther and the seventh season of *Curb Your Enthusiasm* is that one extra step. If avid *Seinfeld* and *Curb* fans are not confused enough already by who's a copy of whom, Larry David has a *Seinfeld* reunion show embedded in the seventh season of *Curb Your Enthusiasm*. But it is not *really* a reunion show, but a *simulation* of a reunion show as Julia Louis-Dreyfus reminds us in HBO's video "*Curb Your Enthusiasm*: It's Not a Reunion Show But It's the Closest You'll Get": "Curb has a tone that is the perfect tone, to do a reunion, a *Seinfeld* show. That is to say, you're not really doing a *Seinfeld* reunion show. You're doing an episode of *Curb*. And that's what this is."[13] And Jason Alexander (who isn't *really* Jason Alexander, but actually Jason Scott Greenspan) states: "It's *Curb*, it's *Seinfeld* . . . we're playing Jason and Julia and Michael but we're not.[14] They're sort of like Larry's idea of them."[15] Jason Alexander, Julia Louis-Dreyfus and Michael Richards come onto the set of *Curb Your Enthusiasm* not only to play their characters, but also play themselves and are brought deeper into the complex layering of simulations in which David exists.[16]

The Show is Unscripted so Therefore Really Real, Right?

Why exactly is *Curb Your Enthusiasm* the perfect show to do a *Seinfeld* reunion show? To start off with, both *Seinfeld* and *Curb Your Enthusiasm* are about "nothing" as *Seinfeld* had previous boasted about in the episode "The Pitch" (0403) when George insists to NBC executives that the sitcom that Jerry and he are writing is about nothing.[17] With *Seinfeld's* plot lines revolving around situations the New York Four get themselves into in their daily lives, mostly taking place in place in Jerry's apartment or in Monk's coffee shop, both shows are supposed to be about nothing and mirror "real life."[18] *Curb Your Enthusiasm* is much of the same, where the plot of the show is strikingly similar every season; director Robert Weide points out in an article for *DGA Quarterly* that viewers can expect this "spoiler" for the next season: "Larry will say and do things that will be misunderstood by people who will get very irritated with him. His troubles will snowball and comedy will ensue until everything blows up in his face."[19]

What makes *Curb Your Enthusiasm* so interesting if it's about nothing just like *Seinfeld* and Larry David is really just George Costanza in a new show? As previously mentioned, there is a studied formlessness about the filming style of *Curb Your Enthusiasm*. Unlike *Seinfeld*, which has a relatively standard sitcom filming style, *Curb Your Enthusiasm* is a hybrid of an observational documentary and a television sitcom. The hybrid, coined as "comedy verité" by critic Brett Michaels, imitates the verité style of filming which viewers recognize from documentaries. When viewers watch a show like *Curb Your Enthusiasm,* they equate the comic verité filming style with the real automatically because of the verité style's association with the real via documentaries.

David's guidelines for the filming of *Curb Your Enthusiasm* are "all location, all in-sequence and no script."[20] Director Robert Weide explains that two handheld cameras are used and the actors and actresses are given relatively short outlines which detail the episode's story arc, but all of the dialogue is completely unscripted and improvised.[21] When asked in an interview if there are a lot of rehearsals before actually shooting, Weide replies,

> You'd think so. No, actually the first time any dialogue comes out of anyone's mouth, the cameras are running. We do a rough camera blocking before we start shooting but I tell the actors just to "blah-blah" the dialogue until the cameras roll. . . You'll see actors—especially Larry—actually laugh in this show. It's not acted laughing and it's not out of character. But when something funny is said that catches you by surprise, the involuntary reaction is to laugh and we capture that because the surprises are real.[22]

The dialogue and reactions appear to be spontaneous and "real" based on this filming style, but Wiede does reveal later in the interview with Sweetzer that often the scenes have to be reshot multiple times to get the take right:

> Sometimes the first take really sings because everything is fresh and no one's trying to replicate what they did before. Sometimes it take[s] six before we even have a clue as to what the hell we're doing. By take eight, we'll have it down pretty well, then I'll go in for coverage. I think the record number of takes was in the thirties, but that was an especially complicated scene. That number includes pick-ups and reaction shots, too.[23]

The final version of an edited episode ends up not really even being close to what is real, as there is no way to tell if scenes are created from the original, spontaneous take described by Weide or if it's the sixth or sixteenth take of a scene. An episode of *Curb Your Enthusiasm* is actually a spliced string of simulated unscripted scenes which presents itself as real through the verité style.

Beyond all of the takes and retakes which further bring a potential real away from itself, *Curb Your Enthusiasm* is one of the most heavily edited television shows on air. The show's two editors, Steve Rasch and Jon Corn, make between three-hundred and four-hundred picture edits for each half hour show. Underneath the picture, they have to make an additional five-hundred audio edits to "massage the dialogue into something clear."[24] Rasch comments that: "'What I'm hired to do is make good sentences . . . The grammar is very important. If it comes in the door and the talk isn't clean, the idea is to patch together good sentences.'"[25] Not only are the takes not really real, the dialogue which was once supposed to simulate reality in its unscripted, improvised manner is then being cut and spliced to simulate some sort of formed dialogue which simulates dialogue in real life. All of the layering of retakes, edits and splicing is mediated through television which is a lens that always already simulates.

Deviating from the "comedy verité" style is the seventh season of *Curb Your Enthusiasm*. The seventh season of *Curb Your Enthusiasm* has the *Seinfeld* show, complete with Seinfeldian sitcom set, embedded in the layers of verité filming. In this season, Larry works to complicate, confuse and play with the idea of doubles of simulations, particularly when he and George interact. The premise of the seventh season is that Larry's wife, Cheryl, has left him and he is trying to win her back by having her star as George's ex-wife, Amanda, on the *Seinfeld* reunion show. In the reunion show, Amanda would get back together with George, and Larry hopes that through the process of working with Cheryl on the *Seinfeld* set, she would get back together with him. In the last episode of the season, "Seinfeld" (0710), Larry misses a date with Cheryl to go over her lines and she invites Jason over instead.[26] Later, Larry catches Cheryl over at Jason's trailer and Jason runs and jumps out the back window when Larry goes to confront him. This is the dialogue of the scene between Larry and Cheryl:

> CHERYL. He's sensitive, he's funny, he's a little neurotic so things like this they get under his skin . . .
> LARRY. What are you kidding! That's George! That's not Jason!
> CHERYL. That's Jason!
> LARRY. That's George! That's all George and that's me! I wrote that stuff. You're not attracted to him, you're attracted to me! I'm George!

Larry is aware that Cheryl is not falling in love with Jason Alexander, but instead with George who is the simulation of Larry, but Larry argues that "He is George!" and further blurs the distinction between himself and George. So what does Larry do to remedy the situation and win back Cheryl? He rewrites the *Seinfeld* reunion show scenes with George and Amanda and has them as physically far apart as possible. When Jason quits after this rewrite, Jerry reminds everyone "you got no George, you got no show . . . it's like a three legged goat" and Larry steps in saying he could play George because "I'm more George than he is."[27]

Is Larry really more George than George or rather more George than Jason Alexander? Julia Louis-Dreyfus comments when asked about Larry David as George: "How's his impersonation of George? It's not good." And even David admits his impression is "just slightly unpleasant." Jason Alexander remarks that it is a "little strange telling the guy that came up with George, how to do George."[28] Even though Larry is supposedly the real George, he cannot be George because the character as simulation has gone beyond Larry to replace him in hyperreality. Though, one has to wonder if the Larry David mediated on any screen, whether it be in *Curb Your Enthusiasm* or an HBO "Behind the Scenes" video, is the real Larry David.

Loss of the Real or Who is the Real Larry David?

As mentioned at earlier in the Canetti quotation, one of the most lucid ways to know the real is lost is recognizing the search to recover it. Baudrillard explains in regards to the search for the real: "The search for proof, indeed the objectivity of the facts does not put an end to this vertigo of interpretation. That is, we are in a logic of simulation, which no longer has anything to do with a logic of facts and the order of reason."[29] In looking for the real in this hyperreal world, we are playing by new rules and the logic is defined by simulations and is unlike any other logic or fact by which we previously functioned.

In searching for the real we only find more simulations; a fetishization of the authentic persists, but it is a "perceived authentic" as the "authentic" is not real, but another simulation. Every pathway taken in the quest for the real only leads to a simulation and the journey is one of confusion as the simulations have collapsed onto each other. But still, people ask and obsess: "Who is the real Larry David? What is he like?" An article in the *New York Times* contains one example of the search for the real as David takes a break after working on a scene of *Curb Your Enthusiasm*: "By the pool of the once-glamorous Roosevelt Hotel on Hollywood Boulevard, he ate his lunch during a break in the incest group scene and subjected himself to questions about the real Larry David. 'I have quite a house,' he said. 'People come over and I go, 'I know, I'm sorry.'''[30] In the search for the real David, one is discouraged to find that he does have a lot of money and a big house, just like in *Curb Your Enthusiasm*. Searchers are not rewarded in finding the real, just a reference and confirmation of the simulation of Larry because there is no "outside" the show anymore.

It isn't just David who is bombarded with questions of the real. Many others from the show, either in the cast or behind the scenes, are constantly interrogated about the reality. Even in a very recent interview in September 2009 which is ten years after the first season, Jeff Garlin, who plays Larry's manager Jeff Greene on the show, is asked about the realness of the character. This example also reveals that it is not just the characters who play themselves (with identical first and last names) who are fetishized, it is everyone. John Kubicek asks in the interview "And what would you say is the biggest difference between your char[acter] and, are you different in personality, or is it mostly just you in the screen?" to which Garlin answers, "What you see in the screen, the personality is pretty close, but the person is not. Because I think that Jeff Greene is a pretty evil guy."[31] Again, there are no true findings of the real, just another dead end and a simulation.

In the epic search for and fetishization of the real in *Curb Your Enthusiasm* there is even an entire book dedicated to trying to trace the real back to its origins. Real outlines of the show are included, pictures of real artifacts from the show, exclusive interviews, commentary on each episode, references and parallels to *Seinfeld*, mannerisms of the characters which are real of the people playing them and most notable—all of the real life events included in the show which really happened to characters, especially Larry David. This book is an homage to the real, although author Deirdre Dolan writes in the introduction of the book: "Bring up Larry David's name and the first thing everyone wants to know is: 'Is he really like that?' Does he really go through life picking fights with guys in wheelchairs, lesbian receptionists, his father-in-law?"[32] She continues to discuss the reason people want to know is because "*Curb* reflects a side of ourselves we've never seen reflected on television—the real side . . . The difference between Larry David on *Curb* and Jerry Seinfeld on *Seinfeld* is that even during *Seinfeld's* cultural reign, no one particularly cared if Jerry was 'really like that.'"[33] While this is all true to a degree, there is something beyond just wanting it to be real because it is a raw side of ourselves. This other unspoken "something" is the fetishization of the real because of the realization that it is lost in a mass of simulations. This is evident in the fact that there is a fetishization to find the "authentic" of many characters of *Curb Your Enthusiasm*. The fetishization also occurred pre-*Curb* in an examination of *Seinfeld* in an essay by Joanna L. Di Mattia titled "Male Anxiety and the Buddy System on *Seinfeld*" which focuses on the fetish of distinguishing the *real Jerry* from the *television Jerry*.[34] There may not have been as much of a fetish to find the real Jerry, but *Seinfeld* may have also aired during a time when the loss of the real may not have been actualized. Regardless now, there definitely is a fetish to find the real, and the characters/actors/actresses of *Seinfeld* are not going to be spared in this search during this post-*Curb* existence.

"Pretty, Pretty, Pretty Good"—Some Concluding Thoughts

Larry David, no matter how hard he tries, cannot transcend the boundary of hyperreality and usurp the role of George which he previously hoped would have partially reclaimed himself as an original. This situation is impossible because there is no original or authentic, just simulations of former originals. David the writer cannot reclaim himself as the real from either of the characters of George or Larry. Not only is David in a multi-layered situation of simulations, but being on television always already creates the context for the process of simulation. People on television, whether acting as a character or "being" themselves, become public figures and exist in a studied media platform which renders them as not "real" anymore in the ontological state of what is considered real. In regards to being a public figure existing in the realm of television, one writer even asked: "Is TV-Being redeemable?"[35] Which in David's case, and other public figures, it appears that it may not be. The reclamation of the self as "real" from the simulational process of the television platform appears to be unattainable.

Although the loss of the real may appear to be something to mourn, for writers like Larry David, simulacra is something which can actively complicate, delight and interest viewers and thinkers as they puzzle through the layers. By recognizing the "loss" of the real we can, in this hyperreal and postmodern world, look at the play which exists via simulations. In the acknowledgement of the complex layers of simulacra, we can better understand the logic, or lack thereof, of our world without necessarily having to try to piece the fragments of reality together again.

Notes

1. For the purposes of clarity, as other contemporaries writing on this subject have done, any reference to Larry David the person will be written as David. A reference to Larry David the character on the show *Curb Your Enthusiasm* will be written as Larry. This is also true for other characters or people as needed.

2. Plato, "The Republic, Book X" in *The Norton Anthology of Theory and Criticism*, ed. Vincent B. Leitch et al. (New York: Norton, 2001), 67–80.

3. Jean Baudrillard, *The Vital Illusion*, ed. Julia Witwer (New York: Columbia University Press, 2000), 61–62.

4. Baudrillard, *Vital Illusion*, 62.

5. Barry Brummett, *The World and How We Describe It: Rhetorics of Reality, Representation, Simulation* (Westport, CT): Praeger, 2003), 14.

6. Jean Baudrillard, *Simulacra and Simulation*, trans. Sheila Faria Glaser (Ann Arbor: University of Michigan Press, 1994), 2.

7. David Lavery and Marc Leverette, "Afterword: Rereading *Seinfeld* after *Curb Your Enthusiasm*" in *Seinfeld, Master of its Domain: Revisiting Television's Greatest Sitcom*, ed. David Lavery and Sara Lewis Dunne (New York: Continuum, 2006), 204.

"The Note" from *Seinfeld: Season 3*, Episode no. 1, first broadcast 18 September 1991 by NBC. Directed by Tom Cherones and written by Larry David and Jerry Seinfeld.

"The Weatherman" from *Curb Your Enthusiasm: The Fourth Season*, Episode no. 3, first broadcast 25 January 2004 by HBO. Directed by Robert B. Weide and written by Larry David.

8. "The Weatherman" from *Curb Your Enthusiasm: The Fourth Season.*

9. "The Pick" from *Seinfeld: Season 4*, Episode no. 13, first broadcast 16 December 1992 by NBC. Directed by Tom Cherones and written by Larry David, Jerry Seinfeld and Marc Jaffe.

10. Baudrillard, *Simulacra and Simulation*, 11.

11. While the idea of a caricature technically presupposes an original, I would argue that what is being caricatured in not an original, but something perceived as an original. The "original" being caricatured in actually a simulation which is considered the real.

12. David Lavery, "Curb Your Enthusiasm" in *The Essential HBO Reader*, ed. Gary R. Edgerton and Jeffrey P. Jones (Lexington: University Press of Kentucky, 2008), 212.

13. "*Curb Your Enthusiasm*: It's Not a Reunion Show But It's the Closest You'll Get," YouTube video, 21:54, from an HBO video posted on 05 Oct. 2009. http://www.youtube.com/watch?v=mofkVvRHT2k

14. Jason Alexander's real name is Jason Scott Greenspan. Jason Alexander is his stage name.

15. "Season 7 Invitation to the Set." HBO video, 2:47, http://www.hbo.com/curb-your-enthusiasm/about/video/invitation-to-set.html

16. Their "pre-Curb" simulation experience can be summed up in this short example: "Julia Louis-Dreyfus? Who's that? Oh . . . you mean Elaine." Julia is no longer Julia Louis-Dreyfus, but is replaced by her simulation Elaine.

17. "The Pitch" from *Seinfeld: Season 4*, Episode no. 3, first broadcast 16 September 1992 by NBC. Directed by Tom Cherones and written by Larry David and Jerry Seinfeld.

18. The location of Jerry's real apartment is not even located in New York City, it is in Los Angeles (a city which is a simulation in itself).

19. Bob Weide, "Funny Business: Stranger than Fiction," *DGA Quarterly*, Fall 2005, http://www.dga.org/news/dgaq_1105/beg_funnybiz-1105.php3?section=news

20. Robert B. Weide, interviewed by Norman Sweetzer, "No Writers, No Script, No Rehearsal: Is This Any Way to Direct a Sitcom?" *Whyaduck Productions*, http://www.duckprods.com/projects/cye/cye-sweetzerinterview.html

21. Ibid.

22. Ibid.

23. Ibid.

24. Deirdre Dolan, *Curb Your Enthusiasm: The Book* (New York: Gotham Books, 2006), 92.

25. Dolan, *Curb Your Enthusiasm: The Book*, 92.

26. "Seinfeld" from *Curb Your Enthusiasm: The Seventh Season*, Episode no. 10, first broadcast 22 November 2009 by HBO. Directed by Jeff Schaffer and Andy Ackerman and written by Larry David.

27. Ibid.

28. "Ep. 70: Larry Playing George." HBO video, 2:47, http://www.hbo.com/#/curb-your-enthusiasm/cast-and-crew/larry-david/video/seinfeld-finale-larry-playing-george.html

29. Baudrillard, *Simulacra and Simulation*, 16.

30. Rick Marin, "The Great And Wonderful Wizard of Odds," *New York Times*. 16 July 2000, http://www.nytimes.com/2000/07/16/style/the-great-and-wonderful-wizard-of-odds.html?scp=1&sq=&pagewanted=1

31. Jeff Garlin, interviewed by John Kubicek, "Exclusive Interview: Jeff Garlin, from 'Curb Your Enthusiasm.'" *Buddy TV*, http://www.buddytv.com/articles/curb-your-enthusiasm/exclusive-interview-jeff-garli-10655.aspx

32. Dolan, *Curb Your Enthusiasm: The Book*, 9.

33. Ibid.

34. Joanna L. Di Mattia, "Male Anxiety and the Buddy System on *Seinfeld*" in *Seinfeld, Master of its Domain: Revisiting Television's Greatest Sitcom*, ed. David Lavery and Sara Lewis Dunne (New York: Continuum International Pub. Group, 2006) 89-107.

35. M.W. Smith, *Reading Simulacra: Fatal Theories for Postmodernity* (Albany: State University of New York Press, 2001), 18.

Chapter 4
Trapped in TV Land: Encountering the Hyperreal in *Supernatural*

Michael Fuchs

The other day, my father—whom I wouldn't refer to as "white trash," since that is too American of a concept for describing a Central-European working class man, but let's call him "Homer Simpsonesque"—asked me what I'm currently working on, a question that he usually avoids at all cost. There I was, staring at his face, wondering what to say or how to approach the issue. Have you ever even considered explaining "hyperreality" or "the fourth order of the simulacrum" to someone who has never heard of postmodern theory (even though there is no singular "postmodern theory")? So, I started talking about how reality has dissolved into hyperreality, the "precession of simulacra," and that postmodernity is characterized by a radical implosion erasing formerly valid differences and boundaries, such as social class and gender. "But that chair I'm sitting on is still a chair, and I'm a man," was the expected reply. Trying to avoid the issue of gender performativity, I responded: "Yes, but the idea of that very chair is influenced by representations of chairs in commercials, catalogues, etc. The chair you're sitting on, anchored in physical reality, and its simulacra are intricately connected." Instead of ignoring the subject and moving on to other things, my dad got loud until I eventually found myself following a rather weird local proverb—"der Gescheitere gibt nach" ("the clever one gives in")—and responding "yeah, sure, you have a point."

The rather inconclusive narrative sketched above, which, for all you know, could be as "real" as the Ecclesiastes quote opening "La Précession des simulacres," doesn't (explicitly) say a lot about popular culture and (hyper)reality—yet. It just so happened that a couple of days later, we were sitting in front of the TV watching an episode of *Supernatural* that had a title whose intertextuality even my father could not miss—"The Song Remains the Same." In the episode, the two main characters, Sam and Dean Winchester, have traveled back in time in order to save their present lives. When contemplating how to approach their parents, who in the present past are younger than Sam and Dean, the following dialogue takes place:

> Dean: What exactly are we gonna march up there and tell 'em?
> Sam: Uhm. The truth?
> Dean: What? That their sons are back from the future to save them from an angel gone Terminator? Oh, come on, those movies haven't even come out yet.[1]

Triumphantly I turned to my dad and said: "See, this is what I'm currently working on. Whatever they tell their parents will shape their parents' idea of who Sam and Dean 'really' are, thus signs, i.e., the words used to describe themselves, precede 'reality' as perceived by their parents. Plus, Sam and Dean are seeing their lives as replicas of movie images." Of course, this is a misuse of Baudrillard's concepts, but good enough an appropriation of his notion that the map precedes the territory in the postmodern age:

> Today, abstraction is no longer that of the map, the double, the mirror, or the concept. Simulation is no longer that of a territory, a referential being, a substance. It is the generation of a real without origin or reality by models: a hyperreal. The territory no longer precedes the map, nor survive it. Now, it is the map that precedes the territory—*precession of simulacra*—it is the map that creates the territory [. . .].[2]

Unimpressed, yet at the same time seemingly shocked by my statement, my father responded: "That's TV. That's not reality!" But is it?

Supernatural Semblance, Simulation, Dissimulation, or . . . ?

As Roland Barthes argues, a fictional work is never expected to present reality, but only to signify reality.[3] This signification of reality is what Barthes refers to as "resemblance" (usually translated as "aesthetic semblance") and is not to be confounded with "simulation," which—not only according to Barthes—is never the goal of any fictional work. Aesthetic semblance becomes particularly interesting when the unrepresentable is signified, as will be briefly discussed below with reference to trauma. Even if the images found in a TV series such as *Supernatural* cannot present 'reality,' they may serve as a medium to access the Real. Already some twenty-five years ago, Umberto Eco noted that by our very nature, we human beings are storytelling animals.[4] As such, (wo)men indirectly access the Real through narratives—we make sense of the world by constructing images of ourselves and the world.

The story that *Supernatural* narrates currently includes 104 episodes (and counting), which is why outlining the plot seems like an impossible task. For readers that don't know the television show, this minimal sketch of the series may be given: the central characters of the show are Sam and Dean Winchester, educated by their father to become hunters, but not your usual hunters; they hunt vampires, demons, ghosts, and all kinds of supernatural entities. The five seasons that have aired to date all feature a season-long story arc[5] woven into and alternating with "monster of the week" episodes. The pre-title sequence of the pilot episode is indicative of the show's structure: the camera shows a night sky with no traces of civilization. A title card indicates the setting: "Lawrence, Kansas; 22 years ago." After some tilting and panning, the camera finally finds evidence of human life—a lonely house. Inside, we see a perfect family: baby Sam, Dean, and their parents, Mary and John Winchester. After a couple of moments, however, the mood abruptly changes; Mary is seen pinned to the ceiling of Sam's nursery, and suddenly she goes up in flames. In the diegetic (as well as "real world") present, Sam's girlfriend Jessica dies just like his mother later on in the episode. Like the opening scene of the pilot, a number of *Supernatural* episodes are set in some small village that seems to be nostalgically out of sync with contemporary technoculture; and when the Winchesters are closer to "civilization," the supernatural appearance of angels and demons makes sure that lights go out or that cell phones have no signals (a contemporary version of one of the defining features of horror—isolation).

Even though small doses of the hyperreal find their way into the U.S.-American hinterlands[6] (e.g., there is hardly an episode in which Sam is not seen surfing the internet), the constructedness of reality doesn't become one of the central tropes of the show until season four, when the show takes a self-referential and meta-textual twist.[7] This first occurs in a rather comic episode, in which the Winchesters find themselves re-enacting scenes of Universal's 1930s horror movies in "Monster Movie" (in black and white, no less). However, "Monster at the End of This Book" takes the question of reality to a totally new, postmodernist level. The episode opens with some seemingly drunk man having visions about Sam and Dean. After some flickering that supposedly indicates a switch from the visualization of the vision to the representation of diegetic "reality," Sam and Dean are seen entering a comic book store. They show their fake FBI IDs and start questioning the owner. After some questions, the owner exclaims: "I knew it! You guys are LARPing, aren't you?"

Dean:	"Excuse me?"
Owner:	"You're fans!"
Sam:	"Fans of what?"
Dean:	"What is LARPing?"

Owner: "As if you don't know. [*confused looks by Sam and Dean*] Live action role playing. And pretty hardcore, too."
Dean: "Sorry, I don't know what you're talking about."
Owner: "You're asking questions like the building's haunted, like those guys from the books . . . what are they called . . . *Supernatural*. Two guys use fake IDs with rock aliases, hunt down demons, ghosts, vampires. What are their names? Steve and Dirk? Sal and Dan?"
Sam: "Sam and Dean?"
Owner: "That's it!"[8]

Sam and Dean learn that their lives have been chronicled in a series of pulp novels, of course entitled *Supernatural*. The brothers decide to talk to Chuck Shurley a.k.a. Carver Edlund, the author of the books, and what follows would likely make John Barth, Donald Barthelme, Kurt Vonnegut, and company proud. I will return to the episode a little later on, but for the moment, let me briefly focus on LARPing. It is well-known that the humanist notion of the subject as a unified self constructed around a stable and fixed identity has been replaced by the postmodern understanding of the subject as fragmented, mutable, and decentered, continually negotiating between several subject positions. While some critics may regard this development as the "'death' of the subject," that is, the end of "the bourgeois monad or ego or individual,"[9] the postmodern subject has rather simply become more elusive. Indeed, according to Judith Butler, "identity is performatively constituted"[10]—the subject only emerges through a performative act. Can there be a more postmodern expression of subjectivity than LARPing? LARPers create a hyperreality in which the participants physically act out their characters' actions, schizophrenically transforming into someone else in the process. It is only appropriate for existence in hyperreality that the shop owner in *Supernatural* thinks of Sam and Dean as LARPers and does not even consider the possibility that the "real" Sam and Dean Winchester are standing in front of him.

Death in Its Plenitude

Unsurprisingly for a show that revolves around the hunting of ghosts, zombies, etc., death takes center stage in *Supernatural*; but how real is death in *Supernatural*, in which ontological boundaries are obviously defined in different ways than in the "real world"? As paradoxical as it may sound, *Supernatural* knows three kinds of death: "real" death, "simulated" death, and what, following Eva Kingsepp, may be referred to as "carnivalesque" death. The latter is the only kind of death that unquestionably ends a person's or some entity's existence—s/he or it is simply no more. What is also characteristic of "carnivalesque" death is the "attempt to capture a notion of death where accentuation of the corporeality of the event, highlighting the bloody, the gory, and the grotesque, is crucial."[11] In *Supernatural*, this kind of death is largely reserved for villains. Whenever the death of one of the "good guys (and girls)" is turned into a visual spectacle, s/he is bound to not return to the show.

The central "carnivalesque" death of the series is Mary Winchester's, which is especially important, for it shows how trauma works in hyperreality. As storytelling animals, humans also mediate trauma through narratives. Increasingly, we experience traumatic events—which occur in physical reality—through mediated representations of these traumatic events. In an attempt to get to the core of the trauma, the Real, we try to separate reality from what Slavoj Žižek calls the "fantasy-frame of reality."[12] However, the result of these attempts is that "the passion for the Real ends up in the pure semblance of the spectacular *effect of the Real*, then, in an exact inversion, the 'postmodern' passion for the semblance ends up in a violent return to the passion for the Real."[13] In extreme cases, this violent return to the passion for the Real leads to people inflicting bodily harm on themselves, as Žižek exemplifies by cutters. In *Supernatural*, the three male Winchesters don't inflict bodily harm on themselves to gain access to the real trauma of Mary's death, but rather willingly expose themselves to all kinds of bodily pain.

Paradoxically, "real" and "simulated" deaths are not that different in certain respects. Both "real" and "simulated" deaths don't lead to the eternal end of someone's (or something's) existence. People that "really" die in the diegetic world of *Supernatural* continue to live in various forms—they may be in heaven, they may be in hell, or they may remain on Earth in a number of shapes and forms (vampires, ghosts, zombies, etc.). However, no matter what post-mortem form really dead people may take, their human existence is over.

Unlike their mother, for the Winchester boys, death is not an end. They may die, but time and again, they are resurrected. Christian symbolism aside, the immortality of the Winchesters is not only indicative of the fact that the longing for eternal life has turned into a cultural obsession in postmodernity, as is evidenced by cosmetic surgeries, cloning, cryonics, etc., but, more importantly, that even death—which could be regarded as the ultimate link to the Real—has lost its link to reality. Indeed, the season five finale even inverts parts of Baudrillard's argument written in response to 9/11: Sam sacrifices himself, wanting to move "the struggle into the symbolic sphere," thus creating "the absolute and irrevocable event"[14]; however, after jumping into a fiery abyss leading into the infernal depths of hell in order to free the world from Lucifer, Sammy is again walking on Planet Earth at the end of the episode. Death has lost its meaning; it is just a simulation of death.

Welcome to the Desert of the Real

After having gone to bartending school and made his tavern more hip and modern, Moe tries to explain postmodernist aesthetics to Homer, Barney, Lenny, and Carl in the *Simpsons* episode "Homer the Moe." "Weird for the sake of weird"[15] is the best he can come up with. Without a doubt, this is a limited perspective on the varied functions of postmodernist arts, and the Baudrillardean mantra of "make believe that the rest is real"[16] cannot be the sole explanation for the use of meta-elements in *Supernatural*. Indeed, the meta-elements in *Supernatural* serve numerous functions, ranging from defining its place in the history of television to smartening[17] the "amateur narratologists"[18] sitting in front of millions of TVs throughout the world. But within the context of reality constructions, it is especially interesting to note that in the diegetic world, the distinctions between "reality" and the supernatural are relatively easily discernible (e.g., when Sam and Dean suddenly find themselves in a TV show that they only watched some minutes earlier in "Changing Channels"[19]); however, this does not mean that diegetic reality and physical reality can be differentiated according to the same rules.

In "Monster at the End of This Book," Sam and Dean confront the author of the *Supernatural* books, which are their lives. When Chuck, the writer, finally accepts that Sam and Dean are neither fans, nor in on a trick played on him, nor hallucinations, he concludes that he must be a god:

> I write things and then they come to life. I'm definitely a god; a cruel, cruel, capricious god. The things I put you through—the physical beatings alone. I killed your father, I burned your mother alive, [*looking at Sam*] and then you had to go through the whole horrific ordeal again. And for what? All for the sake of literary symmetry! I toyed with your lives . . . your emotions . . . for entertainment.[20]

As is revealed in the course of the episode, Chuck is, in fact, not a god but "only" a prophet that chronicles what is to become the Winchester gospel in the near future. In this role, he is, however, only a tool in the hands of the quintessential author-god, the Judeo-Christian god, announced dead by Friedrich Nietzsche in 1882 and by Roland Barthes in 1968.[21] Interestingly enough, the resurrection of the assumedly long-dead author has been noted by scholars with reference to other popular culture texts, too. For example, Terrance R. Lindvall and J. Matthew Melton have commented on the ending of *Duck Amuck* in the following way: in "the world of the simulacra one finds not a referent, but an author."[22] The same may be claimed about the movie *New Nightmare*, at the end of which the main characters' return to "reality" is commented on by the film's creator, Wes Craven, in a voiceover.[23] The crucial difference between the authorial powers in *Duck Amuck*, *New Nightmare*, and *Supernatural* is, of course, that while a religious interpretation is possible in the former two, it cannot be avoided in the latter. The deconstructed structure has rediscovered its center: there is God, there is Truth, and there is a stable and fixed Self—perhaps, to borrow a rhetorical gesture from Derrida. After all, even if God may be there in the world of *Supernatural*, he doesn't care about the events on Planet Earth; even if there is the seeming Truth at the core of which God and his construction of reality feature prominently, contrary to the divine plan, Michael and Lucifer do not face off to decide the fate of the planet at the end of season five; and even if the selves of the central characters of the show seem to be relatively fixed, Sam and Dean in particular know how to play their roles and switch identities as they feel necessary—FBI agent, high school teacher, Lt. Horatio Caine, etc., LARPing in the "real" world, if you will—in short, postmodern subjects par excellence. If one adds their constant "field trips" on the internet as well as their awareness of how their actions are derivative of past figures from the entertainment industry (and the series' self-consciousness as regards its role in the history of TV, its status as cultural product, and its [cyber] fan communities), *Supernatural* is a TV series that wallows in hyperreality; a hyperreality that is, however, not such a nihilistic space as Baudrillard in his master-narrative of postmodern hyperreality—towards which we postmodern subjects feel a certain incredulity—would want us to believe. Indeed, in *Supernatural*, there is meaning hidden beneath the layers of hyperreal pseudo-meaning. As Damien, one of the LARPers at the *Supernatural* convention in "The Real Ghostbusters" puts it: "Our lives suck. But to be Sam and Dean—to wake up every morning and save the world, to have a brother who would die for you . . . well, who wouldn't want that?"[24] In other words, hyperreal images create authentic emotional responses—or, as Umberto Eco may remind us, feelings that we know someone else (should have) felt when s/he was in the same situation. Hyperreality can create illusions, and these illusions keep us from "murdering reality": "As soon as we renounce fiction and illusion, we lose reality itself."[25] Perhaps.

Notes

1. Sera Gamble and Nancy Weiner, "The Song Remains the Same," *Supernatural*, season 5, episode 13, directed by Steve Boyum, aired Feb. 4, 2010 (Burbank, CA: Warner Home Video, 2010), Blu-Ray.

2. Jean Baudrillard, "La Précession des simulacres," in *Simulacres et simulation* (Paris: Galilée, 1981), 10; all translations mine.

3. Roland Barthes, "Le Mythe, aujourd'hui," in *Mythologies* (Paris: Seuil, 1957), 238–240.

4. Umberto Eco, *Postille a Il nome della rosa* (Milano: Bompiani, 1984), 13.

5. In season one, Sam and Dean are looking for their father, who is searching for the murderer of his wife; in season two, the brothers are going after the killer of their mother as they are becoming aware of the fact that they are part of a bigger plan, and Sam discovers supernatural abilities; in season three, Sam and Dean are trying to find a way to free Dean from a Faustian pact; in season four, they are trying to keep demons from freeing Lucifer; and season five (which was initially planned to be the final season, as can be clearly seen from the season finale, but The CW Network decided to renew the series for a sixth season) focuses on their war with Lucifer.

6. Interestingly, the entire series is filmed in Vancouver, BC, and its surroundings, which thus turns into the simulacrum of the United States.

7. It may be added that the reality of the diegetic events is more explicitly questioned as early as season two in episodes such as "Tall Tales," in which parts of what the

viewers visually witness is laid bare as constructed by the Winchester boys as they are reconstructing the recent past as narrator figures.

8. Julie Siege, "The Monster at the End of This Book," *Supernatural*, season 4, episode 18, directed by Mike Rohl, aired Apr. 2, 2009 (Burbank, CA: Warner Home Video, 2010), Blu-Ray.

9. Fredric Jameson, *Postmodernism; or, the Cultural Logic of Late Capitalism* (Durham, NC: Duke UP, 1991), 15.

10. Judith Butler, *Gender Trouble* (New York, NY: Routledge, 1990), 33.

11. Eva Kingsepp, "Fighting Hyperreality With Hyperreality: History and Death in World War II Digital Games," *Games and Culture* 2 (2007): 373.

12. Slavoj Žižek, *Tarrying with the Negative: Kant, Hegel, and the Critique of Ideology* (Durham, NC: Duke UP, 1993), 89.

13. Slavoj Žižek, *Welcome to the Desert of the Real!* (London: Verso, 2002), 9–10.

14. Jean Baudrillard, "L'esprit du terrorisme," published in *Le Monde*, 3 Nov. 2001, accessed 12 Jul. 2010, http://www.lemonde.fr/cgi-bin/ACHATS/acheter.cgi?offre=ARCHIVES&type_item=ART_ARCH_30J&objet_id=677599&clef=ARC-TRK-NC_01.

15. Dana Gould, "Homer the Moe," *The Simpsons*, season 13, episode 3, directed by Jen Kamerman, aired Nov. 18, 2001 (Los Angeles, CA: Twentieth Century Fox Home Entertainment, 2010), Blu-Ray.

16. Baudrillard, "La Précession des simulacres," 26.

17. See Steven B. Johnson's polemical *Everything Bad Is Good for You* (2005), in which he suggests that in the course of the past thirty years, the growing complexity of popular culture has positively influenced U.S. Americans' cognitive development.

18. Jason Mittell, "Narrative Complexity in Contemporary American Television," *The Velvet Light Trap* 58 (2006): 38.

19. Disregarding the extra-compositional implications that are presented in the following, within the diegesis, these constructions are what Baudrillard refers to as 'dissimulation,' i.e., representations in which the differentiation between reality and representations of reality may be masked, but there is still a relatively clear difference between reality and fiction (cf. "La Précession des simulacres," 12). When criticizing the makers of *The Matrix* for misunderstanding his ideas, Baudrillard points out that "either the characters are in the Matrix [. . .] or they are radically outside it [. . .]. But what would be interesting is to show what happens at the juncture of these two worlds. Yet the most embarrassing part of the film is that the new problem posed by simulation is confused with the classical one of illusion, which one could already find in Plato. [. . .] There are no more external Omega points from which to analyze the world, no more antagonistic means" (Lancelin).

20. Julie Siege, "The Monster at the End of This Book."

21. See Friedrich Nietzsche, "Die fröhliche Wissenschaft," in *Friedrich Nietzsche: Werke in drei Bänden*, Vol. 2, ed. Karl Schlechta (Munich: Hanser, 1959), 166–167; see Roland Barthes, "La Mort de l'auteur," in *Le Bruissement de la langue* (Paris: Seuil, 1984).

22. Terrance R. Lindvall and J. Matthew Melton, "Towards a Post-Modern Animated Discourse: Bakhtin, Intertextuality and the Cartoon Revival," in *A Reader in Animation Studies*, ed. Jayne Pilling (London: John Libbey, 1997), 213.

23. See Michael Fuchs, "A Horrific Welcome to the Desert of the Real: Simulacra, Simulations, and Postmodern Horror," in *Landscapes of Postmodernity: Concepts and Paradigms of Critical Theory*, ed. Petra Eckhard, Michael Fuchs, and Walter W. Hölbling (Vienna: LIT Verlag, 2010), 82–85.

24. Eric Kripke, "The Real Ghostbusters," *Supernatural*, season 5, episode 9, directed by James L. Conway, aired Nov. 12, 2009 (Burbank, CA: Warner Home Video, 2010), Blu-Ray.

25. Žižek, *Tarrying with the Negative*, 88.

Part II

The Big Screen: The Simulacra in Contemporary Film

Chapter 5
The Gentleman's Gentleman: The Butler as Simulacra in *The Remains of the Day*

Ann K. McClellan

The concept of servitude is problematic in the United States for several reasons, not the least of which is our history with black slavery. The word "servant" itself seems to signify a permanent under-class, dehumanized, static, bound, attached to the home. It implies a relationship similar to that of "master and slave" that now seems antithetical to the national identity Americans have crafted for themselves post-Civil War and Civil Rights. And yet, servant narratives—on the page, on television, and on the screen—abound in postmodern American culture.

So, why are Americans so interested in stories about British servants when it appears to go against the national identity we have constructed since the 1960s? In *Postmodernism, Or, The Cultural Logic of Late Capitalism*, Fredric Jameson provides one possible answer. He argues that postmodern culture has "the obvious ideological mission of demonstrating, to their own relief, that the new social formation in question no longer obeys the laws of classical capitalism, namely, the primacy of industrial production and the omnipresence of class struggle."[1] However, while democracy's goal may be to convince us that we live "beyond" class structures, those structures continue to influence our society in hidden ways. Perhaps one reason American audiences, in particular, remain interested in servant narratives is because they make visible what we pretend isn't actually there: class structure and struggle. One particular way these servant narratives do this is through authority. I argue that postmodern servant narratives question the traditional master/servant narrative by first reversing the power dynamic in the traditional master-servant dialectic and then by making it impossible for the viewer to distinguish between the two, thus turning the servant into a simulacrum of the master, his class, and authority in capitalist culture.

Perhaps one of the most complex servant narratives under study is Kazuo Ishiguro's *The Remains of the Day*. The main character, butler Mr. Stevens, is near the end of his career, and he spends the novel reflecting back on his life and career, wondering whether he lived a life of "dignity," if he made the right decisions, and whether he will go gone in servant history as one of the "great" butlers of all time. Even though Stevens spends a considerable amount of time in the novel establishing his authority—on butlering, household management, international affairs, and interpreting history—many critics see Stevens as unreliable, most significantly because he constantly revises his narrative and even admits to having a fallible memory. What is so interesting about this unreliability, however, is that while it is made explicit in the novel, it is almost completely absent from the film version, thus turning a book about the unreliability of memory, and the unreliability of the subaltern's speech, into a sincere retelling of a servant's life.[2] In the 1993 Merchant Ivory film version of *The Remains of the Day*, Stevens is portrayed as earnest rather than unreliable, and his master, Lord Darlington, is naively malevolent, leaving the audience without any clear source of class authority. We can no longer rely on the master to lead us in a morally correct direction, as previous narratives have directed; but neither can we rely on the butler to reverse the traditional class hierarchy of master and servant by revealing the "real," behind-the-scenes secrets of the bourgeoisie. Ultimately, in his quest to become the perfect "gentleman's gentleman" in a post-World War II world, Stevens blurs the lines between master and servant until it becomes impossible for the viewer to determine who has the authority to define class.

One thing that makes many Americans uncomfortable with, yet fascinated by, the idea of servants is that they seem to represent a permanent—and more importantly, static—underclass, which goes against American notions of democracy and classlessness. In many British country house narratives of the nineteenth and early twentieth century, for example, it is not uncommon to read about a servant who has been "with the house" for decades, if not his entire life. The opening and closing shots of *The Remains of the Day* reinforce this fear and play with the concepts of exteriors and interiors, important themes that will be discussed in more detail later in the essay. In the first shot of the film, the camera focuses on a drawing of the exterior of Darlington Hall, Stevens' place of residence and employment for over three decades. The scene then shifts to the Darlington Hall auction where Mr. Lewis, the new American owner, proceeds to purchase all of its contents including the butler, Stevens, who is shown in the hall, opening the shutters, implying that while other more transitional items like paintings and furniture can be sold off, the butler remains part of the literal establishment.

The ending of the film similarly reinforces this idea that Stevens is somehow trapped literally by his position within the household. In a scene parallel to the opening of the film, Stevens is once again shown opening shutters and talking about designing a new staff plan. Mr. Lewis and Stevens end up in the old dining room which has now been converted into a game room, with a ping pong table in the center, when a pigeon flies down the chimney into the room. As Mr. Lewis helps the bird escape out the window, the point of view of the camera shifts to the exterior of the house, forcing the viewer to look back at the master and servant inside the house. As the bird/viewer flies away from the window, s/he sees Stevens close the window, watching the bird's flight away from the hall. Ultimately, the camera angle shifts again to the point of view of the bird, flying away from Darlington Hall, looking back down on the building and the man trapped within it. Thus, the opening and closing scenes of the film imply that Stevens has been locked up within the house, a kind of prisoner of his position and class, for decades. Unlike other working class people who looked for work, servants look for "a place," signifying the merging between subject and static object.

Servants represent one locale in particular: England. One thing that seems consistent in many of the popular servant narratives in America is that the domestic classes are British. American shows like *The Fresh Prince of Bel Air* and *The Nanny* include British actors or characters as servants, and PBS's popular series *Upstairs, Downstairs* and *Jeeves and Wooster* are both set in England. P. G. Wodehouse admitted that American expectations significantly affected his portrayal of the famous British butler, Jeeves, who debuted in "Extricating Young Gussie," in *The Saturday Evening Post,* September 1915. Jeeves is a stereotype of Englishness, Wodehouse readily admitted. These "exaggerated dudes" were what Americans wanted; they were what Americans expected the English to be like.[3] In a century of economic crisis and post-imperial decline, the servant narrative has been one commodity that has served the British economy quite well.

Critic Ryan Trimm explains this fascination with British servant narratives as a kind of "national branding" for England. He writes: "The butler is less in service to the State than he is a symbol of the national 'brand', a packaging of the past in the interests of product placement."[4] And this brand, as Carl Freedman points out, is a myth in itself. British servant narratives like *Jeeves and Wooster* and *Upstairs, Downstairs* produce "an ideology of English self-representation both as distinctive postwar cultural form and as highly successful (and economically much needed) British export . . . One of the most successful products of the British culture industry is the myth of Britain itself, or rather of England."[5] This production is, by definition, a simulation of an identity that never existed. Thus, servant narratives allow Americans to buy into a particular kind of nostalgia for a lost imperial identity of their own. By affiliating ourselves with British culture, its structure, and history, Americans gain the authority they lost by breaking away from England two hundred years ago to become a "new" "classless" nation. In their drive to establish their own class structure and authority, America's *nouveau riche* strive to simulate authority by appropriating 'Britishness' through the figure of the British butler.

Part of the "myth of England" that Freedman discusses is the "myth of class" and the employer's status. For example, when author Robert Louis Stevenson became wealthy enough to afford a butler in the late nineteenth century, he wrote to his friends in "mock ecstasy": "We have a butler! . . . He doesn't bustle, but the point of the thing is the style. . . . I shall have it on my tomb: 'He ran a butler.'"[6] Stevenson aptly illustrates the Hegelian master/slave dialectic in his belief that possessing a butler signifies his class position and importance. Hegel argued that the slave mirrors the master's self more through his existence, i.e., presence, than through the work he actually does, and by doing so, reflects back the master's status to himself and to others. The butler, thus, literally becomes a status symbol, a figurative representation—in his very person—of his master's class.[7]

One way that butlers particularly mirror their master's status is through their attire. While household servants like footmen were required to wear livery—ornate uniforms which symbolized the master's wealth to his guests—the butler "colludes with his master's efforts at performing class" by wearing a "costume" much like that of his employer; "exempt from livery, a butler's right to gentlemanly costume was the visible sign of his complicity with his employer's ambitions."[8] By wearing the same "costume" of gentility, the difference between the gentleman and the 'gentleman's gentleman' is elided. What is so interesting about this comment is that while the butler's clothing is seen as a "costume" signifying class, it does not recognize that the master's clothing represents the same thing: a simulacrum of class, status, and identity. Thus, by mirroring the master's class status back to him through his clothing, the butler points out the master's own simulation of class.

In the film of *The Remains of the Day*, the butler Stevens is seen throughout his professional duties in either a suit or tuxedo, just like the master and guests at Darlington Hall. In fact, in some scenes, it is almost impossible to tell the difference between the gentlemen and the "gentleman's gentleman." Midway through the film, for example, Lord Darlington hosts an international conference to discuss Europe's treatment of Germany after the First World War. After dinner, all of the gentlemen go into the parlor to listen to the female German ambassador sing. In the scene, Stevens is shown intermingled with the other male guests, chatting with Mr. Cardinal, Lord Darlignton's godson, and the only way the viewer knows he is a servant and not a guest is because he is holding a wine decanter instead of a glass, like other guests. By portraying Stevens—and all of the other gentlemen—as "suits," the viewer becomes confused. On the one hand, the butler's uniform, like the gentleman's suit, seems to imply a concealed identity underneath, a mythical "origin" of self. On the other hand, the fact that all of the men look the same emphasizes that Stevens—and the men he serves—are all just simulacra, copies of an idea of class, what it looks like, and what it stands for. Stevens is a stereotype, in the psychoanalytic sense. As Homi Bhabha explains, "Like the mirror phase 'the fullness' of the stereotype—its image as identity—is always threatened by 'lack,' and, because it is a form of splitting and multiple belief, the stereotype requires, for its successful signification, a continual and repetitive chain of other stereotypes."[9] If everything becomes surface, if one cannot tell the gentleman from the gentleman's gentleman, then there is no knowable class identity to stabilize society and the servant symbol risks being emptied of meaning.

The butler's "complicity with his master's ambitions"[10] implies that the butler has no desires of his own but rather sublimates his needs to those of his master and, as a result, he becomes alienated from himself. As Michael Terestchenko explains, Stevens' obsession with "dignity" and "greatness" in the film "abolishes the separation between the professional person and the private individual, a split which, in the case of 'merely competent' butlers, cannot but lead to a conscious role-distance when confronted with trials that highlight the fact that they are only playing a part."[11] One way the film illustrates Stevens' split identity is by making him interchangeable with other servant characters within the film, particularly with his father, Mr. Stevens, senior. Early in the film, Stevens informs Lord Darlington that he has hired a new housekeeper, Miss Kenton, and a new under butler. When Lord Darlington asks the new man's name and Stevens replies, "Ah, Stevens, sir," Lord Darlington looks confused. After a hesitation, Lord Darlington states the obvious: "Stevens? [. . .] But that's your name."[12] This comedy of mistaken identity continues when both servants go to exit the room and Lord Darlington asks Stevens for the number of dinner guests that evening and both men turn around in response, thus reinforcing the lack of individualism and sense of self for the butler. In fact, Stevens himself encourages this confusion between himself and his father by insisting that Miss Kenton call his father Mr. Stevens instead of William, his Christian name. As she explains to him, as housekeeper, she has a higher rank than Stevens' father and she reminds him that this duplication in naming may cause confusion among the other servants. Stevens, however, insists. This all comes to a head one evening during the servants' dinner when Miss Kenton informs "Mr. Stevens" that he is needed upstairs and Stevens responds, mistakenly believing she is referencing him, forcing her to explain: "It's for Mr. Stevens, senior, Mr. Stevens."[13] The audience cannot determine whether these scenes are supposed to represent aristocratic attitudes toward servants as mere appendages to the master's will or if they reveal Stevens as a split subject; however, the ways in which the film problematizes ideas of servitude and selfhood make the audience question concepts of interiority, individualism and center.

Just as the previous scene asks the viewer to distinguish which is the "real" Mr. Stevens, the villagers of Moscombe experience similar difficulty distinguishing between the "real" gentleman and his simulacra. It is interesting to note the differences between the novel and film versions here; whereas the novel emphasizes the slippage between the butler and the gentleman in various scenes, the film reduces this down to one exchange between Stevens and the people of Moscombe. Mrs. Taylor, his hostess in Moscombe, informs Stevens, "Dr. Carlisle usually drops in this time of night. He'll be very pleased to meet you. He's a gentleman like yourself."[14] Others question him about his role in politics and international affairs, assuming he is the gentleman he represents, in both the literal and figurative senses. This scene makes the viewer wonder how the citizens of Moscombe know that Stevens is a gentleman. Some characters, particularly those of the working classes, seem particularly persuaded by the trappings of class like clothing, cars, and speech. In this scene particularly, Stevens is dressed in a dated yet distinguished gray tweed suit, most likely a hand-me-down from his former employer, and he is driving an antique Daimler roadster. However, this belief in external identifiers of class is false, since the film makes an effort to point out that the villagers are mistaken in their identification of Stevens as an upper-class gentleman.

So if clothing, cars, residence, and/or speech do not signify class, then it must be something internal and indefinable, yet recognizable. While the villagers may be persuaded that Stevens is the "real" thing—a gentleman—it takes one to know one. In fact, it is only Dr. Carlisle, the other local "gentleman" in town, who is able to recognize Stevens is not who everyone else thinks he is. Driving Stevens back to his stalled car, Carlisle asks, "I say, I hope you don't think me very rude, but, um, you aren't a manservant of some sort, are you?"[15] Carlisle's recognition of Stevens as outside of the gentlemanly class suggests that class is something internal and discernable; one just needs the right credentials to be able to identify it. But then again these class "authorities" are mere country doctors and American *nouveau riche*; the "real" gentlemen like Lord Darlington have been completely discredited by this point in the film due to his alleged traitorous affiliations with the Nazis during World War II. It seems all "authorities" are suspect.

While the role of the servant is to represent the master's status and class position, the very presence of the servant within the household serves as a constant reminder of the "unnaturalness and arbitrariness" of that sign in an upper class world that would like to "present itself as natural, rooted, and fixed."[16] His very identity is a contradiction: "he is a gentleman only insomuch as he is defined in relation to a gentleman of property and social standing; that is, *he is employed to be a gentleman.*"[17] Unlike the aristocratic household where authority can be "proved" by bloodline and inheritance, "the self-made middle classes are always performing, and hence, as spectacle, always under threat of alienation, interrogation, and disintegration."[18] Indeed, "Records of eighteenth-century trials involving adultery among the aristocracy . . . reveal that in the great houses of the English nobility servants were hard to avoid and adept at spying."[19] For example, Wodehouse's Jeeves, although an admitted stereotype and "Americanized" incarnation of servitude, is not an empty figure or without power. In fact, what is most striking about the Wodehouse stories is how much power Jeeves has over other, higher classed characters—including his master, Bertie Wooster. As a child, Wodehouse admitted he had feared the power of butlers: "he used to be in awe of these 'supermen' who 'passed away with Edward the Seventh.'"[20] In fact, Bertie himself doesn't understand why someone with Jeeves' powers of observation, insight, problem solving, and persuasiveness would continue to work as a butler at all: Bertie often wonders "why a man with his [Jeeves'] genius is satisfied to hang around pressing my clothes and what not. If I had half of Jeeves's brain, I should have a stab at being Prime Minister or something."[21] Jeeves consistently has more authority than Bertie, even though Bertie often comments about having to put Jeeves back in his place: "More than once, as I have shown, it has been my painful task to squelch in him a tendency to get uppish and treat the young master as a serf or peon. These are grave defects."[22]

This reversal of power is clearly seen in the publicity photos for the *Jeeves and Wooster* television series. In one excellent mock portrait of Bertie Wooster and his butler, Jeeves is sitting resplendent in a chair, clothed in a three piece black suit that looks more like a modern tuxedo than a servant's uniform.[23] Eyebrows and chin raised, he seems to express simultaneously discomfort in his position at the foot of his master and irony at the seeming reversal of importance. Bertie stands somewhat behind him, a cocky grin on his face, seemingly in on the class "joke," his right hand pulling back his suit jacket and his left behind Jeeves' back on the chair. The image seems to suggest that it is Jeeves who holds the "seat of power" while Bertie waits on him. There couldn't be a better illustration of the reversal between the master-servant dialectic often seen in contemporary interpretations of servant narratives.

Aristocratic authority is similarly under siege in *The Remains of the Day*, as Stevens spends much of his later life defending, obfuscating, and even denying Lord Darlington's Nazi sympathies heading into World War II. Stevens' defense of Lord Darlington seems confusing to many modern viewers. Why remain loyal to a master who sympathized with Nazis and who fired servants because they were Jewish? What they do not understand, however, is the special place a butler had within an aristocratic household. Neither servant nor member of the family, the gentleman's gentleman resided in a place in between traditionally assigned roles. Stevens spends much of his time discussing what it means to be a good butler and how he and his staff should serve their master and his guests. Much of a butler's dignity comes from being of infinite service to his master, even when that service blurs the lines between the professional and overly personal. One example of this in-between status is when Lord Darlington asks Stevens to explain the "birds and the bees" to his adult godson, Mr. Cardinal (played by Hugh Grant in the film). The actual scene in the film, if it wasn't so painful, would be a perfect example of a traditional British comedy of errors. Stevens is clearly uncomfortable with the request, but Lord Darlington persists in insisting that it would be "less awkward" and "easier" for Stevens, a bachelor butler, to explain "the facts of life" than it would be from someone in Cardinal's family or class. Darlington admits to Stevens that his request is "somewhat irregular" and that it "goes beyond the call of duty";[24] however, his request reveals a complex understanding of the complicated position butlers played within households. Clearly, Darlington would not have asked the housekeeper, Miss Kenton, or any of the other servants to take on such an important and personal mission, which shows how close he feels to Stevens. At the same time, however, Stevens is not a human being but a servant, and Darlington doesn't care enough about his feelings of discomfort not to "order" him to talk with Cardinal. The butler's position in-between traditional cultural formations of class, family, and industry challenges and thus deconstructs the boundaries of these very formations from within.

Ultimately, while other servant narratives emphasize the authority of the servant over the master, *The Remains of the Day* gives the contemporary, postmodern democratic reader no such assurances. Ideally, American audiences identify with servant characters in order to reflect a self-image that is democratic and "beyond" class, but audiences often reverse this role to romanticize the past and deny political realities. In particular, we fantasize about British servant narratives because they reflect nostalgia for an older national identity and class structure that we both deny and desire. In our patronage of servant narratives in film, television, and text, we valorize the subversive and transgressive voice of the servant as the "voice of the other," which is symbolic of democratic ideals and identity. What we miss, as viewers, however, is how films like *The Remains of Day* both feed into and challenge these democratic and nostalgic desires by portraying class, through the figure of the butler, as simulacra. Rather than being emptied of meaning, the butler figure thus becomes a symbol of exchange, signifying the elusiveness of class and identity.

Notes

1. Frederick Jameson, *Postmodernism, or, The Cultural Logic of Late Capitalism,* (Durham: Duke UP, 1991), 3.

2. However, there are some critics like Kathleen Wall and Molly Westerman who do not buy into Stevens' unreliability. Rather, they argue that Stephens is a "split subject." Rather than reading his unreliability as moral blindness or "intellectual bias," Wall and Westerman see it as evidence that he is unable to reconcile his private self with public agency.

3. George Watson, "The Birth of Jeeves," *Virginia Quarterly Review* 73.4 (Autumn 1997): 3.

4. Ryan Trimm, "Inside Job: Professionalism and Postimperial Communities in *The Remains of the Day*," *LIT* 16 (2005): 140.

5. Carl Freedman, "England as Ideology: From 'Upstairs, Downstairs' to 'A Room with a View,'" *Cultural Critique* 17 (Winter 1990-1): 79-80.

6. Jean Fernandez, "'Master's Made Away With': Servant Voices and Narrational Politics in *The Strange Case of Dr. Jekyll and Mr. Hyde*," *LIT* 15 (2004): 367.

7. "In these two moments, the master gets his recognition through an other consciousness, for in them the latter affirms itself as unessential, both by working upon the thing, and, on the other hand, by the fact of being dependent on a determinate existence; in neither case can this other get the mastery over existence, and succeed in absolutely negating it. We have thus here this moment of recognition, viz. that the other consciousness cancels itself as self-existent, and, *ipso facto,* itself does what the first does to it. In the same way we have the other moment, that this action on the part of the second is the action proper of the first; for what is done by the bondsman is properly an action on the part of the master. The latter exists only for himself, that is his essential nature; he is the negative power without qualification, a power to which the thing is naught. And he is thus the absolutely essential act in this situation, while the bondsman is not so, he is an unessential activity. But for recognition proper there is needed the moment that what the master does to the other he should also do to himself, and what the bondsman does to himself, he should do to the other also. On that account a form of recognition has arisen that is one-sided and unequal. " G. W. F. Hegel. "Lordship and Bondage," *The Phenomenology of Mind*, accessed 20 July 2010, http://www.marxists.org/reference/archive/hegel/works/ph/phba.htm

8. Fernandez, "'Master's Made Away With,'" 364.

9. Molly Westerman, "Is the Butler Home? Narrative and the Split Subject in *The Remains of the Day*," *Mosaic* 37.3 (2004): 163.

10. Fernandez, ""Master's Made Away With,'" 364.

11. Michael Terestchenko, "Servility and Destructiveness in Kazuo Ishiguro's *The Remains of the Day*," *Partial Answers* 5.1 (2007): 80.

12. Ruth Prawer Jhabvala, *The Remains of the Day*. DVD. Directed by James Ivory. Gloucester, England: Merchant Ivory Productions, 1993.

13. Jhabvala.

14. Jhabvala.

15. Jhabvala.

16. Bruce Robbins, *The Servant's Hand: English Fiction from Below* (Durham: Duke UP, 1993): 18.

17. Karen Scherzinger, "The Butler in (the) Passage: The Liminal Narrative of Kazuo Ishiguro's *The Remains of the Day*," *Liberator* 25.1 (April 2004): 3.

18. Fernandez, ""Master's Made Away With,'" 369.

19. George Watson, "The Silence of the Servants," *Sewanee Review* 103.3 (Summer 1995): 3-4.

20. Eberhard Spath, "P.G. Wodehouse's Jeeves: The Butler as Superman," *Functions of Literature: Essays Presented to Erwin Hoff on his Sixtieth Birthday* 1 (1984): 280.

21. P. G. Wodehouse, "The Artistic Career of Corky," *Carry On, Jeeves* (London: Penguin P, 2000), 38.

22. P. G. Wodehouse, *Right Ho, Jeeves,* (Sitwell, KS: Digireads Pub., 2007), 130.

23. Jeeves and Wooster Portrait, accessed on 10 July 2010, http://images.fanpop.-com/images/image_uploads/J-W-portrait-jeeves-and-wooster461816_962_1462.jpg.

24. Jhabvala.

Chapter 6
The Soft Bodies of Arnold Schwarzenegger

Randy Laist

Arnold Schwarzenegger's early film career takes place among the company of a number of other stars such as Sylvester Stallone, Chuck Norris, and Lou Ferrigno who represent what Susan Jeffords called the 80s "hard body," figures who personify a Reagan-era faith in the indomitable power of the white heterosexual male. In 80s blockbusters such as *Rambo: First Blood Part II* (1985), *Top Gun* (1986), and *Die Hard* (1988), the protagonists act as emblems of reliability, pitting the sheer massiveness of their physique against various forces of cruelty and degeneracy that are typically diminutive, effete, and foreign. The hard body action hero kicks ass not only in the name of American white male hegemony, but, even more fundamentally, for the sake of the reality principle itself. The hardness of the hard body is akin to the hardness of the rock which Dr. Johnson kicked in refutation of Bishop Berkley's idealist philosophy. According to Jeffords' analysis, the hard-body heroes revise Hollywood mythologies of cowboys and adventurers by stripping fantasy narratives of their fantastic atmosphere and asserting their stark reality. "Their greatest act of heroism, as it turns out, is in believing that not only their heroic predecessors but they themselves are real."[1] The hard body action film provides a metanarrative that consolidates power in the hands of white men while simultaneously anchoring reality itself to a stable set of familiar coordinates. Arnold was, as Jeffords called him, "the hardest of the hard bodies,"[2] and most of his movies from the mid-80s, including *Commando* (1985), *Raw Deal* (1986), and *Predator* (1987), strictly adhere to the hard body formula. As the foremost emblem of the hard body genre, Arnold employed his hyperreal persona in the service of an anti-postmodern cinematic statement: that reality is real and it will kick your ass.

This statement comes across with unique directness in *The Running Man* (1987). *The Running Man* depicts a society in which a fascistic government uses television as a means of distracting the citizenry, disposing of political dissidents, and disseminating misinformation. In attempting to replace social reality with highly controlled television images, the government of this future aspires to create a false reality and prop it up as the real reality. This is a familiar style of social control that we have seen in classic dystopian narratives from *Brave New World* to *Fahrenheit 451*. As always is the case in such narratives, the evil government is resisted by a force that represents the truth and authenticity which the government is committed to suppressing. Early in the movie, Arnold falls in with a band of rebels who constantly plot to block the government's television signal and "broadcast the truth." The movie is structured around the differentiation between the illusion represented by the government and the reality represented by the rebels. The government-edited footage of Ben Richards opening fire from a helicopter on a crowd of civilians is a deliberate distortion of the raw footage that reveals that Richards actually refused to obey the order to open fire on the civilians. When Arnold bursts into the television studio at the end of the movie, he violently refutes the televisual illusion of the show by vanquishing it with the undeniable forcefulness of Reality Itself, which he brings with him not only in the form of the "true" footage of what happened in the helicopter, but, even more fundamentally, in the brutal poetry of his physical "hard body" appearance.

In the 90s, however, Schwarzenegger, having achieved unparalleled success as a box office draw, turned to a series of roles which seem intentionally selected to explore the relationship between reality and illusion in a more nuanced way. In the same way that Clint Eastwood began interrogating his own ultraviolent "Dirty Harry" persona in *Unforgiven* (1990), Schwarzenegger's roles in the early 90s indicate a curve toward self-consciousness, irony, and destabilization of simplistic moral dichotomies of right and wrong, as well as of ontological dichotomies of real and unreal. In doing so, Schwarzenegger and Eastwood are both part of a wider cultural trend that is evident throughout American cinema in the 1990s. Between the collapse of the Berlin Wall in 1989 and the terrorist attacks of September 11, 2001, American cinema is very noticeably characterized by a mood of Baudrillardian hyperreality epitomized by popular films such as *JFK* (1991), *The Player* (1992), *Lawnmower Man* (1992), *Jurassic Park* (1993), *Natural Born Killers* (1994), *Pulp Fiction* (1994), *Forrest Gump* (1994), *12 Monkeys* (1995), *Scream* (1996), *Face/Off* (1997), *Pleasantville* (1998), *The Truman Show* (1998), *Being John Malkovich* (1999), *The Matrix* (1999), *Existenz* (1999), *South Park* (1999), *Memento* (2000), and a vast catalogue of similar interrogations into the slippery relationship between existence and representation. Considered within this context, Arnold's four major films of the decade, *Total Recall* (1990), *Terminator 2* (1991), *The Last Action Hero* (1993), and

True Lies (1994), are key texts in this substantial genre of 90s hyperreal cinema and Arnold himself can be clearly identified as a central figure in the hyperrealization of the American imagination. The heroes of these movies can be described as "soft bodies," in the sense that Baudrillard speaks of "soft technologies" or Sontag describes photography as "soft murder." "Soft Bodies" are not what they are; they are saturated with illusory qualities. They may appear dreadful and imposing; they may loom as massively as Arnold Schwarzenegger, but when you shift your perspective, you realize that the soft body is actually made up of dreams and optical illusions. The soft body hero is not a force of nature, because he's not quite natural. Unlike conventional protagonists in Western literature, the soft body hero is not a poem to the noble virtues of humanism, because the soft body is never quite human. When we follow the story of the soft body hero, we are not hanging on the suspense of an existential threat (Will the character survive or not?), but of an ontological threat (Is the character alive or not? Does the character exist or not?).

If *The Running Man* affirms the distinction between reality and illusion, *Total Recall* dramatizes the implosion of any such distinction and definitively establishes Arnold's revamped ontological identity as an icon of hyperreality. In Paul Verhoeven's film, Schwarzenegger plays Quaid, a construction worker in the not-too-distant future who is obsessed with the fantasy of going to Mars. So compelling is this fantasy that he seeks out the services of an artificial memory implantation service, Rekall, to buy a simulacral memory of a Martian vacation. Throughout the rest of the movie, the audience is never certain whether what we are watching is taking place in reality or in Quaid's mind. The fact that the details of Quaid's journey seem to correspond with the details of the Rekall package Quaid purchased leads us to conclude that the narrative is in fact a dream, but the possibility that Quaid may have chosen those details out of a subconscious memory of his real Martian experiences pushes us back to the possibility that Quaid really is and has been the secret agent Hauser. The drop of sweat on Dr. Edgemar's forehead is enough to convince Quaid that the doctor is lying when he tells Quaid that his experiences are all in his head, but as soon as Quaid shoots Dr. Edgemar, the walls of the room explode in an apparent confirmation of Edgmar's prediction that if Quaid opted to pull the trigger, "The walls of reality will come crashing down." Ultimately, the ontological mood of *Total Recall* never collapses into either realism or fantasy but remains suspended in a register that is the essence of Baudrillard's fourth stage in which the image "has no relation to any reality whatsoever."[3]

As the central and perhaps sole inhabitant of this new kind of reality, Quaid is a unique type of hero. When we root for him to escape danger, defeat Cohaagen, win the girl, and save Mars, we do so alongside an awareness that the danger, Cohaagen, the girl, and Mars itself are all possibly nonexistent. To be sure, this is always part of the experience of the consumer of fiction, but *Total Recall* pulls the issue of the suspension of disbelief into the texture of the story itself to depict a world in which lived experience is infused with the ontological quality of fiction. Quaid himself has to keep his own doubts at bay in order to act, and it is not until his very last line in the movie that he can voice his "terrible thought" that all of his accomplishments have been illusory. But Quaid is in a unique position to be able to cope with this vertiginous atmosphere of porous reality due to the fact that his own psychology is a hyperreal construct. If his adventures on Mars are illusory, then he is really himself, and his actions, although they are imaginary, are at least rooted in a real past of true experiences. But if we assume, as Quaid does, that his Martian adventure is real, then "Quaid" is a fictional consciousness programmed by Cohaagen as part of an elaborate plan to crush the Martian rebellion. Quaid is really Houser, a confederate of Cohaagen who has had his memory erased and replaced with the patsy Quaid, who will be deleted once Cohaagen's plan has come to fruition. Throughout the movie, while Quaid thought he was behaving heroically, he was really acting out Cohaagen's script. Quaid's hyperreal environment redounds upon himself; his own psychology has always already been a component of his simulacral context. It's not that he has been advancing Cohaagen's agenda (real) while he thought he was utilizing his own free will (illusory). Rather, it is the case that Quaid's free will itself has been engineered by Cohaagen. Quaid's psychology is not controlled or manipulated; rather, it is colonized on a genetic level. Under such circumstances, political agency becomes either self-defeating or absurd. If Quaid does manage to overthrow Cohaagen's regime of totalitarian control, it is due to the fact that the influences of hyperreality cannot be contained. Quaid may be a fictional construct, but he is a fictional construct come to life with a determination to push his narrative through to the end. When Quaid defeats Cohagan and reinvents Mars, it is a case of the fiction murdering the reality and establishing its own kingdom. The landscape of terraformed Mars that Quaid surveys at the end of the movie is saturated with an ontological ambiguity that reflects his own ambiguous existential status.

In *Terminator 2*, the most successful movie of his career, Arnold reprises his iconic hard body role as a killing machine from the future. In the sequel, however, Arnold's cyborg body is reconfigured as ontologically soft. There is no ambiguity concerning the ontology of the first Terminator. Its steely endoskeleton epitomizes the being of the hard body reduced to its essential qualities: ruthless efficiency in pursuit of a single goal. In the first movie, there is no question about the reality of the Terminator. It is here to kill you and that is all there is to know about it. Its lethality is a brutal indicator of its reality. As an envoy from a world of artificially intelligent machines, the Terminator defines the essence of technology as inhuman and hostile. Sarah Connor's fight with the Terminator is a preliminary skirmish in what the movie tells us will be the total war of human beings against the machines in the decades to come. Naturally, we side with the real human beings against the artificial machines. Arnold's Terminator robot appears to be human, but this is a deception. Beneath his disguise, he is clearly a false imitation. The metaphysical border between real and illusory is upheld throughout *Terminator 1* in a very deliberate way. When Arnold returns for the second movie, however, he has been reprogrammed as a guardian angel for the main human characters. He is still a robot and, what's more, he is still the same make and model of robot from the first movie, but his relation to the real/illusory binary has become more complex. Whereas the first movie dramatized a vary stark contrast between the real human beings and the cybernetic imitation, the presence in *T2* of a second robot that is even more exotic than Arnold's model creates a continuum of humanity, with Arnold occupying an ambiguous middle-ground between the flesh and blood human beings with whom Arnold's robot develops a relationship and the extremely alien T-1000 model of Terminator. The T-1000, a product of cutting-edge CGI technology, makes Arnold's android seem quaintly human by comparison. Compared to this strange entity, Arnold's robot's physiology is reassuring in its more conventional limitations and vulnerabilities. Arnold might not be human, but he is "more human" than the T-1000. In addition, John flips a switch inside Arnold's skull that allows the Terminator to learn. As the Arnold robot begins to pick up pieces of slang and comes to understand the meaning of emotional pain, the movie suggests that human beings do not have a monopoly on the capacity for human behavior. Although he is a machine, Arnold has access to qualities that we are most likely to associate with human psychology. The division between the real humans and the cybernetic impostors has become blurry, and now in order to survive, Sarah Connor must overcome the knee-jerk technophobia that she learned from the first movie and ally herself with a world of simulacra.

In addition to his role of softening the ontological boundary between human and non-human, the robot played by Arnold in *Terminator 2* also embodies a temporal paradox that confounds classical temporality. Baudrillard defined the essence of hyperreality as "the generation by models of a real without origin or reality."[4] Rather than a cause followed by an effect, or an original reality followed by a representation, hyperreality is a condition in which copies have abandoned all reference to an originary reality. In his analysis of time in Baudrillard's hyperreality, William Bogard explains that "simulation is like a miracle, for only a miraculous technology could revive the past in the present (and project it endlessly into the future) . . . The miracle produces a cyborg time-traveler."[5] Arnold's robot, of course, is Baudrillard's cyborg time-traveler come to cinematic life. Not only does he embody Baudrillard's temporal paradox in his status as a robot that has not been invented yet, but *T2* takes this hyperreal condition a step further when Arnold and his human friends break into the Skynet laboratory where the Arnold robot was invented and destroy all the research that led to his existence, effectively terminating Arnold's unborn self. According to *Back to the Future* rules, Arnold should disappear from existence at this point, but Arnold's variety of existence does not rely on past circumstances. Like Quaid, who is told by Kuato that he is defined entirely by what he does in the present rather than by his past or his memories, Arnold's *T2* robot existentially log-rolls on a free-floating temporal precession in which copies need have no reference to any past original. To make matters even more complex, the research that Arnold and his friends destroy in the Skynet laboratory is itself based entirely on a piece of the Terminator robot from the future that Sarah had destroyed in 1984. If the Terminator machine had never been sent back in time in the first place, the Terminator machine and all the planetary robotic infrastructure that dominates the post-apocalyptic future of the Terminator mythology would never have come into existence. The Arnold machine gave birth to itself in a closed temporal loop that is outside of linear time and in which the copy is literally its own original. Moreover, this hyperreal entity is the perfect foster-father for John Connor, who, as a representative of the new techno-savvy generation, is himself a hyperreal construct, having orchestrated the circumstances of his own conception by sending Reese back in time to impregnate his mother. Both the machine and the boy are hyperreal, detached from the old binaries of nature/technology, past/future, and authentic/artificial. Naturally, as John's mother comes to realize, the hyperreal robot is the perfect father for her hyperreal son. "Of all the would-be fathers who came and went over the years, this thing—this machine—was the only one who measured up. In an insane world, it was the sanest choice." The brave new world of hyperreality compels Sarah to reassess her twentieth-century prejudices about the relationship between reality and simulation.

In *The Last Action Hero*, Schwarzenegger again depicts a character whose artificiality makes him an ideal father-figure. The film tells the story of Danny Madigan's travels between the real world and the fictional world of his favorite "hard body" movie franchise. The differences between Danny's world and the world of *Jack Slater* are played for laughs and the movie's end restores the integrity of the barrier between the two worlds, suggesting that *The Last Action Hero* is meant to be a fundamentally conservative reaffirmation of the separation of reality and unreality. But at the same time, as in Pirandello's *Six Characters in Search of an Author*, the very attempt to communicate the difference between reality and fiction ultimately discloses the various correspondences between them. When Danny enters the fictional world, he discovers that fictional characters have real emotions, even as Danny, as a real person, has an emotional life that is centered around the fictional events of the movies he loves. When Slater enters the real world, he actually functions very effectively there, impressing Danny's mother as a viable husband and father-figure, and bending the laws of physics and physiology, which seem just as malleable in the real world as they are in the screen world, to heroic purposes. The climax of the movie takes place on opening night of the new *Jack Slater* film, where the bad guys who have escaped from the fictional world plan to assassinate Arnold Schwarzenegger as a way of getting rid of Jack Slater. Schwarzenegger appears as himself, with his wife Maria Shriver and a handful of other celebrities doing cameos of themselves, a spectacle which has the effect of folding the outside world of media celebrity into the fictional world of the film, provoking in the audience a sense that they are shuttling back and forth between fiction and reality in much the same way as Danny. A journalist interviewing Arnold opines that Schwarzenegger's own incredible biography could only be possible "in Hollywood"; Schwarzenegger himself, as a "real" person, is a product of the quasi-fictional space of "the movies." The fictional Arnold saves the real Arnold from an assassination attempt, compounding the sense that reality and fiction are existentially codependent. Although the Berlin Wall of the screen's surface is reestablished between Jack and Danny at the end of the story, the movie has promoted the impression that this barrier is not as impermeable as we had taken it to be.

The title of the movie *True Lies* seems intentionally to echo the definition of hyperreality that has been associated with Umberto Eco: "the authentic fake." If the more conservative screenplay of *The Last Action Hero* attempts to restore a separation between artifice and reality, *True Lies* derives its narrative momentum from the dismantling of such a separation. As in *The Last Action Hero*, Schwarzenegger plays a double role as both a fictional character and a real person. In one life, Harry Tasker is a cartoonishly efficient super-spy, and in the other life, he is a "real-life" suburban husband and father. The twist is that they are both the same person, and the story moves toward the integration of the fantastic and quotidian profiles of contemporary existence. If Harry is more successful and more fulfilled in his fictional life than in his real life, it is only a symptom of a broader epidemic of fictional values and fictional characters insinuating themselves into the home and usurping its living occupants. Harry's partner explains that Harry's failure to raise his daughter is mitigated by the fact that he is not her "real" father anyway; "her parents are Axl Rose and Madonna. You can't compete with that." Fictional characters have already de-realized the structure of the nuclear family. The only way that Harry can compete with fictional rivals to his patriarchal authority is to become a fictional character himself. Fortunately for Harry, he already is a fictional character, so all he has to do to win back his daughter's love is to rescue her from the roof of a skyscraper with a Harrier jet in a breathtaking special effects sequence. *True Lies* specifically contextualizes the "reality crisis" on the domestic level of daily experience within the geopolitical situation of a post-Cold War world characterized by loose nukes from former Soviet republics coupled with the emerging threat of Arab terrorism. Whereas the Cold War provided a stable framework for maintaining rigid distinctions between values such as us and them or reality and unreality, the new situation reflects a viral reorganization of the territoriality of danger. The threat is now on the home front, which is transformed into a possible site of duplicity and betrayal. The movie's solution to exorcising the home of this stultifying atmosphere of suspicion is to reorganize the family unit as a hyperreal construct. Trust issues between husband and wife are neutralized by the reinscription of both partners into the fictional space of the action-adventure genre. They are still a "real" family—they sit around the dining room table and thumb-wrestle with their daughter—but they slip seamlessly into their alternative lives as a team of international spies. Their new sense of domestic security is accomplished through their shared immersion in a hyperreal existence.

The terrorist attacks of 2001 marked an abrupt end to the speculative atmosphere of hyperreality that had dominated American cinema throughout the previous decade. The fact that the distributors of the Schwarzenegger vehicle *Collateral Damage* (2002) chose to delay its release out of sensitivity to the September 11 attacks is a kind of metaphor for the manner in which the new decade jolted Arnold himself out of his cinematic existence and into historical-political reality as "The Governator" of California. In his transition from personifying hyperreality in a cinematic mode to personifying it in a political mode, Arnold seems to have achieved the feat accomplished by Jack Slater in *The Last Action Hero*, striding between ontological registers in such a way as to require us to reassess our naïve assumptions about the relationship between reality and representation.

Notes

1. Susan Jeffords. *Hard Bodies: Hollywood Masculinity in the Reagan Era.* (New Brunswick: Rutgers University Press, 1994), 63.
2. Jeffords, 141.
3. Jean Baudrillard. *Simulacra and Simulation.* Trans. Sheila Faria Glaser. (Ann Arbor: University of Michigan Press, 1994), 6.
4. Baudrillard, 1.
5. William Bogard. "Baudrillard, Time, and the End." In *Baudrillard: A Critical Reader*. Ed. Douglas Kellner. (Oxford: Blackwell, 1994). 318-9.

Chapter 7
Jesus in Hyperreality: Baudrillard and Mel Gibson's *The Passion of the Christ*

Amarnath Amarasingam

The passion scenes in the life of Jesus have been depicted often in paintings, narrative, and film. These depictions have not always been tame, but they have equally never been as gory and drawn out as the nine-minute and forty-second brutal beating of Christ portrayed in Mel Gibson's *The Passion of the Christ.*[1] Gibson's *Passion* is definitely not akin to Fra Angelico's "Descent from the Cross" (ca. 1430) in which Christ has very few markings on his body. Looking at the painting, one is more consumed by the triumphant Christ than the suffering one. For Angelico, Christ's wounds "elicit our pity, but don't confront us with horror. And the world surrounding his Christ is a bright, luminous, spring day, that reminds us of the rebirth of creation itself."[2] Lovis Corinth's painting "The Red Christ" (1922), on the other hand, is more similar to Gibson's gruesome depiction. Corinth's painting is dark and bloody; a soldier is seen to the left of Christ offering him a sponge to quench his thirst while a soldier on the right thrusts his spear into Jesus' side. The painting is dark, splotchy, and dreadful, reminding us more of death and suffering than of rebirth.

Corinth and Angelico's works are both paintings, however, and so force the imagination to do much of the interpreting. If we stare at Corinth's painting for nine minutes and forty seconds, the atmosphere of the painting will produce images different from the ones we would get if we devoted the same amount of time to Angelico's "Descent from the Cross." Films often do the imaginative work for us, and Gibson's production may be a quintessential example of this. This paper is little concerned with questions of anti-Semitism in the film, and with whether such gruesome violence is in keeping with Catholic or Protestant theology. The former has been exhaustively dealt with elsewhere, and the answer to the latter question largely depends on the type of theology with which one enters the theater. This paper explores how the relationship between consumerism and Baudrillardian hyperreality in the United States influences the way in which films, especially religious films or films dealing with scripture, are received by the viewing public. It is argued that in the process of depicting the passion story of Jesus through film, Gibson manufactures a commodity that is subsequently "bought" by viewers. At the same time, in a hyperreal environment, such simulated commodities become "more real than real," and could, in essence, precede any reading of scripture. Thus, by using *The Passion of the Christ* as a case study, I argue that films depicting religious subjects in general, and scriptural narratives in particular, have enormous influence over the ways in which religious themes are incorporated into one's religious life and the manner in which scripture is read thereafter. Through examining several viewer responses to *The Passion*, this paper attempts to understand the influence of the film on the religious life of individuals.

Baudrillard: Simulation and Hyperreality

In the earlier phase of Jean Baudrillard's (1929-2007) career, he was primarily focused on shedding light on modern consumer culture, and what it meant for contemporary society. His first book, *The System of Objects*, explores the culture of mass consumption that characterizes the contemporary world. As Best and Kellner note, "The project operates within the framework of a subject-object dialectic where the subject faces a world of objects which attract, fascinate, and sometimes control the individual's perception, thought, and behavior."[3] The themes explored in *The System of Objects*—how subjects experience, dominate, and get dominated by objects and signs in everyday life—can be found in much of Baudrillard's subsequent writings.

It is his later work, from the late 1970s onwards, that he is most famous for, and which is most pertinent for this paper. Building on Marshall McLuhan's insights on the medium and the message, Baudrillard argued that both "are in the process of merging to the point where our lives are becoming media-created simulations of reality."[4] For Baudrillard, the era of modernity which has been dominated by industrial capitalism and production has come to a close and been replaced by a new era of "simulations" and "hyperreality." Baudrillard begins his famous essay "The Precession of Simulacra" with the story by Jorge Luis Borges in which imperial cartographers "draw up a map so detailed that it ends up covering the territory exactly."[5] In the Borges fable, the territory nevertheless precedes the map, and determines the characteristics of the map. Baudrillard argues that the Borges fable is an example of second-order simulacra, which, for him, lasted from the Industrial Revolution until the middle of the twentieth century. This was the period of mass production where copies are produced based on an original blueprint. Today, however, the Borges fable has been surpassed, and third-order simulacra have taken its place. In third-order simulacra, all that exists are copies without originals—the map precedes the territory and "it is the generation by models of a real without origin or reality: a hyperreal."[6] Signs come to pervade the world, and determine the nature of social experience. Following McLuhan's lead, Baudrillard argues that the boundaries between simulation and reality have disappeared in the postmodern world, and simulations have effectively replaced reality. As Best and Kellner point out, "Simulation devours the real and, like a Cheshire cat, leaves behind nothing but commutating signs, self-referring simulacra that feign a relation to an obsolete real."[7] Any value that may have existed in the difference between the two has disappeared. As Baudrillard argues, it is in the difference between the representation and the reality that "poetry" is found, and, in contemporary society, the poetry is dead or dying. As Mann notes, in today's world, "Our television 'friends' seem more alive to us than their flesh-and-blood equivalents. We communicate by email and relate to video-game characters like Lara Croft better than to our own friends and family."[8]

It should be clear, then, that simulations are not simply fakery; they are not pretending to be reality. As Baudrillard points out, "Whoever fakes an illness can simply stay in bed and make everyone believe he is ill. Whoever simulates an illness produces in himself some of the symptoms."[9] Arguing that reality is no longer real, Baudrillard interprets Disneyland as a soothing construct that exists to convince us that reality is actually real. As he states, "Disneyland is presented as imaginary in order to make us believe that the rest is real . . . it is no longer a question of a false representation of reality (ideology) but of concealing the fact that the real is no longer real."[10]

Baudrillard's notions of simulation and hyperreality are useful in examining how films, or more specifically, films about scripture, serve to provide a sense of realness for adherents of a certain faith. The scenes portrayed on the screen often become more real than what is contained in scripture or tradition. They become manufactured, free-floating products that have sway over how religion is subsequently experienced. Relating to film and other elements of American popular culture as commodities is a fundamental element of contemporary consumer society. Since religious groups, including Christian communities around the world, "inevitably borrow much of their worldview, structures, and practices from the culture around them," they are also influenced by consumer society.[11] As Miller humorously notes, the fact that some Christians do not eat meat on Fridays depends on the culture that surrounds them. In a vegetarian society, these Christians would be unique not by the fact that they refrained from eating meat on Fridays but that they consumed meat every day except Friday. As Miller states, "Consumerism joins the many other cultural dynamisms—nationalisms, ethnic identities, colonial domination, and class divisions—that have encumbered and encompassed Christian communities' attempts to live their faith and continue to do so."[12]

Religious individuals in attempting to "live their faith," therefore, are influenced by the secular media and the ways in which it portrays religion. As Loomis has pointed out, to view the relationship between religion and the media as conflictual is misleading and unproductive. As he states, "Religion and the media should not be viewed as polar opposites; the relation is more likely to be a symbiotic one that can be understood as having more to do with what a person perceives and experiences while receiving a message than with what the message looked like to begin with."[13] In other words, the religious significance of a media product can best be established in the ways in which it is received by religious audiences and incorporated into their religious lives.

Surrounded by consumerism, Christian and non-Christian communities alike interact with the outside world through "habits of interpretation and use" that inadvertently treat "cultural objects as consumable decoration."[14] In other words, American consumerism is not solely driven by values such as greed, materialism, and selfishness. Even denunciations of such principles "are such attractive values that they easily become commodities in themselves."[15] What is at work seems to be functioning at a deeper level. Consumerism, fundamentally, is a way of seeing the world, and interacting with it. Religious beliefs and symbols, like toys, become "objects of consumption," and Gibson's film is one of the best examples of how this functions.[16] I also argue that the observations of Loomis and Miller, among others, can best be contextualized within a Baudrillardian understanding of the contemporary world as one where the boundaries between the real and the unreal are increasingly blurry.

Mel Gibson's *The Passion of the Christ*

The Passion of the Christ was released in 2004 to widespread anticipation and intrigue among many circles, both religious and secular. It has since become one of the most popular and profitable films of all time. Working with a production budget of thirty million dollars, it has since grossed over six hundred million dollars worldwide. Even prior to the release of the film, controversies arose surrounding whether the film was historical, anti-Semitic, too violent, and whether it was being portrayed objectively in the media.[17] It seems that Gibson went to great lengths to ensure that the film would align closely with scripture. For example, Gibson had religious leaders and experts view the film, had Jesus speak in Aramaic, etc. In addition to scripture, elements of the film were constructed through the use of other religious writings and traditions.

I begin first by exploring the process of commodity construction in the making of *The Passion* before turning to how it was consumed by some Christian viewers. *The Passion* does indeed adhere to the broad strokes of the scriptural narrative. However, Gibson has also taken the liberty of letting his own imagination fill in the gaps. As Jenkins points out, *The Passion* "is somewhat unique in that it takes a discrete section of the various Gospels and conflates them along with material from outside the Gospels."[18] In other words, Gibson has actively manufactured a product that is to be consumed by those who view the film. This, of course, is perfectly allowed since it is in fact an artistic endeavor. But, in a hyperreal world, religious films and scripture, which Baudrillard would argue are equally simulated, compete in the same arena, and carry equally strong claims to truth. The gospels provide only terse descriptions of what happened to Jesus immediately after his arrest. For example, Matthew 27:26 and Mark 15:15 state that "after flogging Jesus" Pilate handed him over to be crucified. Mark 16:19 states that the soldiers struck his head with a reed and spat on him. In Luke 23:16, Pilate tells the Jewish authority that Jesus has done nothing to deserve death and "I will therefore have him flogged and release him." John 19:1-3 states that "Pilate took Jesus and had him flogged" and that the soldiers kept mocking him and "striking him on his face." The scriptural passion narrative alone could not have provided Gibson with enough descriptive material to paint the narrative that he did.

Gibson, however, maintains that he is not straying far from scripture. As he told the Roman Catholic international news service, Zenit, "We've done the research. I'm telling the story as the Bible tells it. I think the story, as it really happened, speaks for itself. The Gospel is a complete script, and that's what we're filming."[19] As one reviewer countered, "while the film is partly based on the gospels, it reflects far more the traditional Catholic iconography of Stations of the Cross" where Jesus is "scourged and crowned with thorns."[20] However, as Craig Evans notes, "If we blend together all of the details of the four New Testament Gospels, [only] five of the first ten Stations of the Cross are attested: (1) Pilate's hall of judgement, where Jesus is condemned to death; (2) the place where Jesus receives his cross; (5) the place where Simon of Cyrene is compelled to carry the cross; (8) the place where Jesus meets the women of Jerusalem and utters his warning, and (10) the place of crucifixion."[21] There is no mention in the Gospels of Jesus falling three times, meeting his mother, or having his face wiped by a woman.

More than this, the film draws from another source, Catherine Emmerich, an early nineteenth-century Augustinian nun, mystic, and stigmatic. Gibson extensively used an account of her visions told to and written by poet Klemens Brentano called *The Dolorous Passion*. As J.D. Crossan states, "Her book forms the basis of the script for his film and, indeed, casting could be done from some of her character descriptions."[22] The scourging scene especially is taken right from Emmerich, who "describes how over the period of three quarters of an hour 'ruthless soldiers' whipped Jesus so ferociously that they 'tore off large pieces of flesh' and 'penetrated to the bone,' leaving his body 'perfectly torn to pieces,'"[23] The scant descriptions found in the gospels are expanded, and made more horrific simply by the use of Emmerich's visions and worsened, no doubt, through the Gibson's interpretation. Thus, through the appropriation of several sources and Gibson's own interpretation of each, a commodity is constructed, and a specific *type* of Jesus is promoted.

The primary focus of the film as stated by Gibson and several subsequent reviewers is to refocus the Catholic faith on what is important: the passion narrative. As Robert Johnston argued, "Both church and Hollywood have sanitized the story, downplaying the full horror and thus the full glory of Jesus's suffering and death on the cross."[24] For Johnston, *The Passion* is just what is needed to realign attention to Jesus' suffering. Many believe that the ethical teachings of Jesus, the focus on the Golden Rule, and his life example have replaced his saving grace. The focus has shifted away from the man who, in the words of Emmerich, "fell on his face, overwhelmed with unspeakable sorrow, [when] all the sins of the whole world displayed themselves before him."[25] Gibson is apparently attempting to overturn H. Richard Niebuhr's criticism that Christians now believe that "a God without wrath brought men without sin into a kingdom without judgement through the ministration of a Christ without a cross."[26]

Gibson wants Christians, particularly Catholics, to recognize the suffering of Christ, and, through the film, rebukes "the Christian public at large who might also be tempted to turn away in order to avoid confronting the intensity of Jesus' suffering."[27] *Plugged In Online*, a publication of James Dobson's Focus on the Family, argues similarly that, "Christians with a sanitized view of biblical history—or complacent believers who've come to take Christ's suffering for granted as they've matured in the Lord—will be shocked into a new appreciation of his divine sacrifice."[28] As Gibson noted in an interview: "I think we have gotten too used to seeing pretty crosses on the wall, and we forget what really happened . . . But when you finally see it and understand what he went through, it makes you feel not only compassion, but also a debt."[29] But, "what really happened" is simply a manufactured commodity that Gibson puts forth as reality. The scourging scene, lasting nearly ten minutes, is easily the most powerful and disturbing in the film. However, large parts of it were taken from the visions of an early nineteenth-century mystic. In a world of simulations, Emmerich's visions are placed alongside scripture as equally viable and as an equal testimony to "what really happened."

Responses to the film have varied greatly, ranging from the *Passion* being called part of "High Renaissance tradition" to being branded as appalling and revolting.[30] The film quickly produced what media researchers call an "active audience," who interacts with the material through interpretation or protest.[31] Academics, media personalities, film reviewers, and individual viewers all had strong reactions to the film. Acclaimed critic Roger Ebert pointed out that at least one-hundred minutes of the 126-minute film "are concerned specifically and graphically with the details of the torture and death of Jesus. This is the most violent film I have ever seen. . . You must be prepared for whippings, flayings, beatings, the crunch of bones, the agony of screams, the cruelty of the sadistic centurions, the rivulets of blood that crisscross every inch of Jesus' body."[32] Another commentator, A.O. Scott, stated in a review that the Christian viewer is drawn into a "sadomasochistic paradox" in which the faith in Christ and his suffering forces the viewer to abandon "the ordinary human response," which is essentially wanting "for the carnage to stop."[33] *Plugged In Online* similarly notes that, "Those who have chosen to follow Christ will experience a bizarre emotional paradox while viewing the brutality. Each blow to the face, lash with the whip and nail through his flesh is simultaneously repellent and indisputable testimony of divine love."[34] Andy Rooney, on *60 Minutes*, called Gibson a "wacko" and, on the other end of the spectrum, critic Stef Loy stated, "Never before has the language of cinema had the potential to challenge the church at large to wake up to the reality of film. It is here to speak and move, to challenge our preconceived notions about life, to affect us in ways that no other medium will ever aspire to."[35] It could be added that no other medium has the potential to be used by the Church for purposes of proselytization. As Billy Graham stated, the film "is a lifetime of sermons in one movie."[36] Indeed, Gibson hoped that *The Passion* would be used for evangelism. As he notes, "The Holy Ghost was working through me on this film, and I was just directing traffic. I hope the film has the power to evangelize . . . everyone who worked on this movie was changed. There were agnostics and Muslims on set converting to Christianity."[37] In the hyperreal world, everything competes in the same playing field. The nightly news is as simulated as a film and a film is as simulated as a statement *ex cathedra*. We see this collapse of boundaries in the debate over *The Da Vinci Code* as well as the *Passion*, where fiction and film are debated as if they are non-fiction and documentary.

The most profound collective action came from Evangelical communities in the United States, who have often protested the cultural output of Hollywood. Many conservative Christians embraced the film and were inspired "to book whole theaters and to share it as religious ritual."[38] Evangelical film review websites often brutally criticize violence on the screen. A review of *Fight Club* stated that after the school shooting at Columbine, the movie was "completely irresponsible" and called it "one of the most dangerous movies released" that year.[39] *Plugged in Online* calls *Fight Club* a "dangerous Hollywood head trip" that "could inspire machismo among distraught males convinced they have nothing to lose."[40] The same kind of negative review was launched at films like *Scream 2* and *Natural Born Killers*. These same groups praised *The Passion* although it contained much more violence than films like *Scream 2*. The *Passion* did not receive the same criticism from evangelical groups for depicting such brutal violence.

One reviewer brushed aside the violence contained in the film as acceptable because it was true. Another stated, "*The Passion of the Christ* is rated R for sequences of graphic violence . . . [but] I personally believe that this time the 'R' should stand for 'Redemptive.'"[41] *Plugged In Online* similarly noted that, "unlike films with lesser ratings that exploit brutality, it doesn't at all glamorize maliciousness and murder. Indeed, it is disturbing for all the *right* reasons."[42]

Another viewer posted this in response to a Christian Answers film review: "I watched as my Lord and Saviour was scourged, beaten, spat upon, and crucified for me. Tears rolled down my cheeks in rivers . . . I am closer to Jesus now after witnessing his sacrifice for me in it's [*sic*] full horror and brutality."[43] The above statement is not referring to representation, but to a real "witnessing," and as such we see an example of Gibson's "map" preceding the "territory." The passion of Christ will likely be filtered through *The Passion of the Christ* for many of these viewers.

Some viewer responses posted on The Independent Movie Database (IMDb) are indicative of the deep emotional reaction that many individuals had: "I took three people with me last week on Easter and one of them prayed with me and was saved"; "It made me want to go to church again"; "If you are a believer, this movie at least should convict you to fall in love with the Savior and walk with him daily."[44] The brutal violence in the film was not deterring or revolting for many believing viewers—it was something that they needed to watch, "painful and difficult though that may be, because it is part of the same calculus of salvation, 'necessary' for them to view just as much as for Jesus to have suffered."[45] Many viewers recalled feeling a sense of deep guilt and shame either because they had forgotten the sacrifice Jesus had made for them, or because they continue to sin while knowing all about it. As one viewer wrote on *Tribute.ca*:[46]

> "It really made me think and look at life from a whole other point of view. Mel did an awesome job and made people believe in God. I had lost faith in him until I saw this movie . . . Amazing . . . I've never cried during a movie before this one"

Another noted that the film had renewed his/her relationship with Jesus:

> "I had very little faith in God and Jesus Christ, but this movie moved me so much that I have begun the journey to bringing Christ back into my life. When the movie ended, no one moved, everyone sat there in silence for a moment, all you could hear were sniffles and sobbing. This movie will move you. It is violent, but it only depicts the reality of what happened"

Another individual stated that the film had made him rethink the path his life was on and encouraged him to be a better person:

> "Last night I was on the net for hours after seeing the movie trying to learn about Jesus. I cried and cried and as a guy I am not supposed to do this. I feel like a changed person and I've made my first call to a drug rehab and I will become a better person. Thanks Mel, your film gave me hope of a better life."

Another important reason that the *Passion* was so successful, especially with Christian audiences, was the trouble Gibson went through in filming it, and, after its release, in dealing with criticisms. Many hindrances to the production of the film such as Jim Caviezel (who played Jesus) getting hit by lightning on more than one occasion all served to enliven Gibson and his evangelical supporters that they were engaged in a metaphysical battle with dark forces. Such perceived mythical struggles helped to underpin the symbolic value of the commodity soon to be available. After the release, critics of the film were attacked as if they were critiquing Christianity itself. As one reviewer stated, "To many people, it's a masterpiece and saying anything bad about the film becomes akin to saying something bad about Christianity."[47] Websites such as www.SeeThePassion.com arose and millions of Christians flocked to the theaters. They felt Christianity was under attack by the cultural Left in the United States, and went to see the film as a collective obligation and a religiously defensive maneuver.

Conclusion

Hyperreality, for Baudrillard, is the blurring of the real and the unreal. When the real is no longer simply in front of the viewer, such as staring at a landscape, but is rather produced artificially through technological manipulation, these synthetic models come to replace reality. These synthetic models can include, for example, magazines touting ideal bodies, ideal sex, ideal men and women, and ideal lifestyles. These ideals are not "unreal" but, for Baudrillard, are hyperreal, realer-than-real. The same can be argued for the world of TV where the boundaries between reality and fiction are blurred. Thus, soap opera actors may encounter an angry individual on the street who derides them for something their character did on the show. Raymond Burr, who played lawyer Perry Mason, received thousands of letters yearly asking for legal advice. Similarly, *The Los Angeles Times* asked whether *Law and Order* star Fred Thompson's run for President of the United States in 2008 would be affected by an obscure role he played as an anti-Semitic, white supremacist almost nineteen years ago. As Best and Kellner state, in the world of simulation, "models become more real than the actual institutions, and not only is it increasingly difficult to distinguish between simulations and reality, but the reality of simulation becomes the criterion of the real itself."[48]

Take for example the interesting tidbit of information from the shooting of *The Passion*: Mel Gibson ordered a $350,000 robot to be crucified instead of actor Jim Caviezel. "A bitter wind on the set meant it was dangerously chilly to hang Caviezel on the cross in just a loincloth—so the amazingly lifelike model was put together by special effects crew on the set and operated by Gibson via a remote control."[49] Millions of viewers, then, were having deep emotional reactions to the movements of a robot, while believing this was "what really happened." As Joe Egan states, "technological advances have had such a significant impact on human imagination and perception that representational images of events can to a large extent overshadow the events themselves."[50] It matters little, therefore, that scenes from *The Passion* are manufactured from several sources; these scenes are now reality, a simulated reality, more real than what may have really happened. As the supportive viewer and reviewer remarks above make clear, the film produced in individuals a deep sense of guilt at their own sin and a deep appreciation of what Jesus did for them. As tears flowed down their cheeks, they reaffirmed the reality of the unreal and infused with meaning the remote controlled movements of a Jesus writhing with pain and sacrifice.

Notes

1. The Passion of the Christ. Dir. Mel Gibson. Perf. Jim Caviezel. 2004. DVD. Warner Home Video, 2004.

2. Mark A. Wrathall. "Seeing the World Made New: Depictions of the Passion and Christian Life." *Mel Gibson's Passion and Philosophy*. Ed. Jorge Garcia. (Chicago: Open Court, 2004): 10.

3. Steven Best and Douglas Kellner. *Postmodern Theory*. (New York: The Guilford Press, 1991): 113.

4. Douglas Mann. *Understanding Society: A Survey of Modern Social Theory*. (Toronto: Oxford University Press, 2008): 283.

5. Jean Baudrillard. *Simulacra and Simulation*. Sheila Faria Glaser (Trans.). (Ann Arbor: University of Michigan Press, 1994): 1.

6. Ibid.

7. Steven Best and Douglas Kellner. *The Postmodern Turn*. (New York: The Guilford Press, 1997): 100-101.

8. Douglas Mann. *Understanding Society: A Survey of Modern Social Theory*. (Toronto: Oxford University Press, 2008): 284.

9. Jean Baudrillard. *Simulacra and Simulation*. Sheila Faria Glaser (Trans.). (Ann Arbor: University of Michigan Press, 1994): 3.

10. Ibid. 12-13.

11. Vincent Miller. *Consuming Religion: Christian Faith and Practice in a Consumer Culture*. (New York: Continuum, 2005): 25.

12. Ibid.

13. Kenneth Loomis. "Spiritual Students and Secular Media." *Journal of Media and Religion*. 3.3 (2004): 151.

14. Vincent Miller. *Consuming Religion: Christian Faith and Practice in a Consumer Culture*. (New York: Continuum, 2005): 25.

15. Ibid. 19.

16. Ibid.

17. William Brown, John Keeler and Terence Lindvall. "Audience Responses to *The Passion of the Christ*." *Journal of Religion and Media*. 6.2 (2007): 88.

18. Jenkins, Bill. "Jesus Christ, Superstar? Why the Gospels Don't Make Good Movies." *Journal of Religion and Film*. 12.2 (October 2008): http://www.unomaha.edu/jrf/vol12no2/Jenkins_Passion.htm

19. Gibson quoted in Mark Silk, "Gibson's Passion: A Case Study in Media Manipulation?" *Journal of Religion and Film*. 8.1 (February 2004): http://www.unomaha.edu/-jrf/2004Symposium/Silk.htm

20. O'Brien quoted in Joe Egan. *Brave Heart of Jesus: Mel Gibson's Postmodern Way of the Cross*. (Dublin: The Columbia Press, 2004): 54.

21. Craig A. Evans. "The Procession and the Crucifixion." *Jesus and Mel Gibson's* The Passion of the Christ. Eds. Kathleen Corley and Robert Webb. (London: Continuum, 2004): 131.

22. John Dominic Crossan. "Hymn to a Savage God." *Jesus and Mel Gibson's* The Passion of the Christ. Eds. Kathleen Corley and Robert Webb. (London: Continuum, 2004): 17.

23. Clodagh M. Weldon. "The Passion Re-Cut: If 'It is as it was', Why the Redaction?" *Journal of Religion and Film*. 10.2 (October 2006): http://www.unomaha.edu/jrf/Vol10No2/Weldon_PassionRecut.htm

24. Robert K. Johnston. "The *Passion* as Dynamic Icon: A Theological Reflection." *Re-viewing the Passion: Mel Gibson's Film and Its Critics*. Ed. S. Brent Plate. (New York: Palgrave, 2004): 56.

25. Emmerich quoted in John Dominic Crossan. "Hymn to a Savage God." *Jesus and Mel Gibson's* The Passion of the Christ. Eds. Kathleen Corley and Robert Webb. (London: Continuum, 2004): 24.

26. H. Richard Niebuhr. *The Kingdom of God in America*. (New York: Harper and Row, 1959): 193.

27. Mark A. Wrathall. "Seeing the World Made New: Depictions of the Passion and Christian Life." *Mel Gibson's Passion and Philosophy*. Ed. Jorge Garcia. (Chicago: Open Court, 2004): 10.

28. "The Passion of the Christ." *Plugged In Online*. June 9, 2008. http://www.pluggedinonline.com/movies/movies/a0001657.cfm

29. Gibson quoted in Mark A. Wrathall. "Seeing the World Made New: Depictions of the Passion and Christian Life." *Mel Gibson's Passion and Philosophy*. Ed. Jorge Garcia. (Chicago: Open Court, 2004): 10.

30. Joe Egan. *Brave Heart of Jesus: Mel Gibson's Postmodern Way of the Cross*. (Dublin: The Columbia Press, 2004): 55.

31. David Croteau and Williams Hoynes. *Media Society: Industries, Images and Audiences*. (London: Pine Forge Press, 2003): 266-269.

32. Roger Ebert. "The Passion of the Christ." Rogerebert.com: Movie Reviews. 24 February 2004. 15 November 2008. http://rogerebert.suntimes.com/apps/pbcs.dll/-article?AID=/20040224/REVIEWS/402240301/1023

33. A.O. Scott quoted in Mark A. Wrathall. "Seeing the World Made New: Depictions of the Passion and Christian Life." *Mel Gibson's Passion and Philosophy*. Ed. Jorge Garcia. (Chicago: Open Court, 2004): 12.

34. "The Passion of the Christ." *Plugged In Online*. June 9, 2008. http://www.pluggedinonline.com/movies/movies/a0001657.cfm

35. Stef Loy quoted in Robert H. Woods, Michael Jindra and Jason Baker. "The Audience Responds to The Passion of the Christ." *Re-viewing the Passion: Mel Gibson's Film and Its Critics*. Ed. S. Brent Plate. (New York: Palgrave, 2004): 166.

36. Billy Graham quoted in Joe Egan. *Brave Heart of Jesus: Mel Gibson's Postmodern Way of the Cross*. (Dublin: The Columbia Press, 2004): 51.

37. Mel Gibson quoted in "The Passion of the Christ." *Plugged In Online*. June 9, 2008 http://www.pluggedinonline.com/movies/movies/a0001657.cfm

38. Neal King. "Truth at Last: Evangelical Communities Embrace *The Passion of the Christ.*" *Re-viewing the Passion: Mel Gibson's Film and Its Critics*. Ed. S. Brent Plate. (New York: Palgrave, 2004): 151.

39. Ibid. 154.

40. "Fight Club." *Plugged In Online*. June 9, 2008 http://www.pluggedinonline.-com/-movies/movies/a0000363.cfm

41. Quoted in Neal King. "Truth at Last: Evangelical Communities Embrace *The Passion of the Christ.*" *Re-viewing the Passion: Mel Gibson's Film and Its Critics*. Ed. S. Brent Plate. (New York: Palgrave, 2004): 156.

42. "The Passion of the Christ." *Plugged In Online*. June 9, 2008. http://www.pluggedinonline.com/movies/movies/a0001657.cfm

43. Quoted in Neal King. "Truth at Last: Evangelical Communities Embrace *The Passion of the Christ.*" *Re-viewing the Passion: Mel Gibson's Film and Its Critics*. Ed. S. Brent Plate. (New York: Palgrave, 2004): 157.

44. Quoted in Kelly Denton-Borhaug. "A Bloodthirsty Salvation: Behind the Popular Polarized Reaction to Gibson's *The Passion.*" *Journal of Religion and Film*. 9.1 (April 2005): http://www.unomaha.edu/jrf/Vol9No1/DentonBorhaugBloodthirsty.htm

45. Ibid.

46. "The Passion of the Christ Movie Reviews." *Tribute.ca*. June 9, 2008 http://www.tribute.ca/reviews/The+Passion+of+the+Christ/7945

47. Quoted in Kelly Denton-Borhaug. "A Bloodthirsty Salvation: Behind the Popular Polarized Reaction to Gibson's *The Passion.*" *Journal of Religion and Film*. 9.1 (April 2005): http://www.unomaha.edu/jrf/Vol9No1/DentonBorhaugBloodthirsty.htm

48. Steven Best and Douglas Kellner. *Postmodern Theory*. (New York: The Guilford Press, 1991): 120.

49. Pisa quoted in Joe Egan. *Brave Heart of Jesus: Mel Gibson's Postmodern Way of the Cross*. (Dublin: The Columbia Press, 2004): 69.

50. Ibid.

Chapter 8

Making Love, Making Violence Avatar-Style: Hyperreal Connection and Deconstructive Identification

Amanda LeBlanc

Avatars are non-theistic motifs of representation now integral to North American electronic culture. Though they are putatively judged unreal by their very nature as mechanical projections of the Self, it is difficult to ignore a bustling industry of simulated entertainment wherein human forms are parodied, morphed, protracted and dismantled to denote progress. One jarring effect of this evolution is how avatars work to trump humanity—a utopian subject position that is tenuously discrete from technic agencies. Hyperreality forces connections between users and figures that are more-than yet oddly less-than human, Facebook personas and cinematic aliens, for instance; as such this discussion probes assertions of how we encounter identity in popular cultures ever more defined by technological reproduction. I use deconstruction to claim simulated shapes operate as both obstacles and passageways to dangers and pleasures of the real. Suspicious of "teletechnology," Jacques Derrida writes little on the subject compared to his work on written language.[1] Nevertheless he affords symbols of recognition and exclusion that prove useful in my analysis: *hymen* mechanics along with *autoimmunity* functions anticipate the tendency for avataristic bodies to collide in spaces of late capitalist media, which are often deemed violent and sexual.[2] Underpinning my philosophical discussion are the material, political and sociological implications of avatars, particularly if they manifest in the way Emily Apter describes, as "prosthetic extension[s] of desubjectivated agenc[ies]" rather than plainly as digital doubles of the Self.[3]

The chief role of avatars, arguably, is to sustain users through perils of hyperreality. In this postmodern realm, according to Jean Baudrillard, reality is already a product of replication so it "becomes the allegory of death, but it also draws strength from its own destruction, becoming the real for its own sake, a fetishism of the lost object."[4] Avatars appear to be reproductive vessels of lost affect, the sort of constructs Fredric Jameson may have vetted in his seminal Marxist critique of postmodernism. This being said, their prophylactic character solicits pleasure as a mode of constant possibility. My assumption, here, is that affect is ambivalent: it encapsulates a search for original impressions—sensory, emotive and intellectual—within the simulacrum, where only derivations can be found,[5] as much as it carries real prospects of the virtual in converting affect and sensation from one modality to another.[6] Avatars, through their junctions of human and machine qualities, signify the "in-between-ness" of Derrida's *hymen*, his barrier paradigm, which stands for neither purity nor reproduction, "desire nor pleasure . . . future nor present." Strangely the hymen is all of these things, none of these things and in-between these things at once, insofar as it destabilizes how the terms "outside" and "inside" are structurally separated and distinguished from each other. Following Derridean thought each avatar, then, is a "textile" spun from and within constitutive gaps between mechanical and organic identities.[7] Picking up on normative biological terms, furthermore, Derrida states the hymen is a woman's "protective screen" that confounds the inertia of virginity with the rupture of coitus[8]—consequently one finds troubling the belief that simulation precludes intimacy insofar as it contrasts dead avatars with their living users. What we have is not a simple opposition. The fulfillment proffered by avatar culture, however counterintuitive, must be considered if not valued beside views of media as a juggernaut of distraction.[9] Avatars work to shield us from ourselves by exposing us to the many vicissitudes of the techno-world. They promulgate fantasies concerning individual and collective discovery. They tell us we can be and know anyone or anything in newly global arenas. These mottos should not be dismissed simply because they deal in "falsity," for they are not necessarily or categorically false.

To describe how the freedom purported by avatars comes with great risk of contradiction and loss, I bring into dialogue two recent specimens of hyperreality: Facebook and James Cameron's 3-D American film, *Avatar* (2009). Both set new standards in linking twenty-first century consumers with their technologies. Whilst Facebook grips the Internet via social ubiquity, *Avatar* helms North America's movie industry as the most sophisticated, profitable and ostensibly arousing viewing experience to date. Most significantly, each phenomenon reconceptualizes taxonomies of intimacy in agreement with violent strides of new media. To quote Baudrillard, "we kill images with meaning."[10]

Let us first discuss Facebook as a commercial meeting place for users, or avatars. Servicing upwards of 500 million users, including "entrepreneurs and developers from more than 190 countries,"[11] this networking site aims to revitalize word-laden Email scripts with visually active agents offering multiple points of access across the globe. Users post 2-D spectacles of their unions: immediate status updates, actions ranging from rough to amorous, videos and stills of everyday life. Everything occurs on the virtual wall. In hymeneal fashion, the wall is open and closed to all who accept, ignore and refuse friend requests—it is a semi-permeable surface, really an annexation of each user's body, by which we parley familiarity in virtual fashion. The opportunity to forge bonds quickly becomes an injunction when Facebook recommends new friends and politely scolds us for losing track of old ones. Perhaps more unsettling than reuniting with long lost flames, who return to us with an uncanny force, is feeling the power of pseudo-strangers, who appear on our Home pages or News Feed pages every day. We may not be able to physically recognize them on the street and off the grid; nevertheless these alien bodies help determine our popularity as avatars, and they help constitute frameworks for perceiving selfhood. The more friends we amass, the more, to use Pierre Bourdieu's now widespread terms, *social* and *cultural capital* we command. Lisa Nakamura maintains, "When users create or choose avatars on the Internet, they are choosing to visually signify online in ways that must result in a new organization and distribution of visual cultural capital."[12] Avatar profits are noteworthy—according to a 2007 appraisal, virtual microcosms contributed approximately US $28.15 billion to the gross domestic product.[13] So if our status gains momentum, we are more saliently courted by online consumerism: not only can friends buy us birthday smooches, marketing companies can take advantage of open yet opaque privacy settings, initiated as default functions in 2009, to harvest our data. Given the frenetic pace at which Facebook adds applications, quizzes, games, etc., privacy policymakers everywhere are scrambling to keep up. Therefore to read Facebook personas as avatars is to foreground thrilling, risky trade offs between "real" and "virtual" relationships.

We are told freedom is social superfluity, imagining life beyond a day-to-day cohort, beyond constraints of material necessity within our neighborhoods, work places and geographical, affective locations. We are also afforded subjective liberties of the masquerade. Users are free to download any legal image for their profile picture, or make their own avatar at websites like mywebface.com and imvu.com. The film *Avatar* is so iconic it has spawned its own genus of post-human branding, an idea I expand upon later. Reface.me churns out blue alien visages that boast the impact of skin color as a performative device.[14] Nakamura, in *Digitizing Race*, argues avatars are not so much alter or trans egos for public display and approval; analogous to how I will adopt Derrida's textile of ego (de)formation, Nakamura takes an intersectional approach to the analysis of specific digital imagery by threading together "diasporic, ambivalent" utterances of "nation, race, gender, and age" for increasingly "closed," taste-driven markets.[15] Emily Apter makes a similar claim to the avatar as a "partly subjectified, partly impersonalized vicissitude of the [Freudian-Lacanian] drive." Apter posits the avatar "is no fixed entity," forming "an equalized playing field of egoic drives and aims that undercuts fantasies of omnipotence and possessive individualism."[16] Could any one avatar incarnate a specialized multitude of cultural, consumer, sexual and political desire? Seemingly, their protean quality bears witness to the "philosophical antinomy" Jameson invokes when discussing political and utopian implications of the Internet: he asks, "does it relate or does it separate and disperse? Is it the sign of identity or of multiplicity?"[17] Facebook negotiates certain singularities, assigning those who wish to remain incognito "neutral" silhouettes (a white form with a spike of hair, or this same form with a shoulder-length, flipped do—two vaguely gendered, racialized looks) then prompting these incognito users to find their picture or *face*. Furthermore, Facebook, like any other social media apparatus, manages ambiguities of representation by systematically asking users to disclose and thus re-confirm the organizational import of identity markers like age, biological sex, sexual orientation and relationship status.

Post-structural grammar might inform social networking ideology. Though Derrida distinguishes medial communication from writing, claiming that the former tricks us into forgetting the simulacrum by relentlessly projecting an immediate presence (while the latter does not), there is room to appreciate how new media, like language, can enable and constrain the human body.[18] Perhaps there is nothing outside the digital text in the same way that nothing escapes linguistic mediation.[19] We need not decry the superficialities of Facebook in contrast to "naturally" human counterpoints, which do not stand on their own, but only as part of the discursive configuration called *différance*. On the one hand, gaining a legion of cyber-buddies requires, to a certain degree, surrogate associations bereft of the kind of touching one achieves when meeting friends in person. Could we physically meet *all* of our cyberfriends? This might be an exhausting and pointless exercise. On the other hand, the fact that social media is pervasively used to communicate meaning about public and personal identity tells us that Facebook is not exactly a game-changer; it is symptomatic of our media-saturated consciousness. To put it another way, Facebook is a product and producer of what we have come to expect of social relations. When using social media, we encounter an already mediated principle of identification, and we continue this deferral process as we accept cyberfriends three or four times removed from the face-to-face encounter.

In "You Have 0 Friends,"[20] a 2010 episode of America's hit animated series *South Park*, cultural satirists Matt Stone and Trey Parker claim Facebook regimens make for co-dependent avatars who postpone useful relations by wasting each other's time with minutiae of the Internet. Mainly unknown townsfolk pepper schoolboy Stan Marsh with requests as soon as he enters the social grid, e.g., fertilizing digital crops in *FarmVille*. Most egregious is the way Stan's father needles him to "Poke Grandma!" and reconfirm family relations, now threatened, under the proviso of electronic compliance. Because *South Park*'s cachet relies on the naïve politics of children who are sager than most adults, it seems Stan contests his own reification as an avatar by quitting Facebook, hence leaving a system he deems inane and void of real cultural capital—a rather idealistic concept. Such a repudiation of hyperreal friendship sets Stan apart from another child, the unpopular Kip, who is happily cloistered in his room, in *FarmVille*, even as his parents believe him to be outside physically enjoying a friend's farm.[21] Stan may be a young target for discourses of techno progress; however, according to Parker and Stone, he possesses enough common sense to observe a Facebook friend for what it is: "some kind of commodity for a person's status." Our dizzying aggregate of occasions to interface presents said fears about people being objectified and overexposed electronically, about users facing unstable convergences between selfhood and, to borrow Apter's term, "avatarity." Apter remarks, "the avatar has become the face of it-ness, who-ness, and what-ness mediating community and unseating the subject's . . . self-possession."[22] Ergo, before any opting out can occur in the South Park example, mechanical resolve must be asserted as Stan discovers Facebook has the ability to materially subsume him, however temporarily. In a battle scene of hyperreality *par excellence*, one that inquires "Who is more powerful: the user or the profile?,"[23] Stan plays and wins a game of Yahtzee against his avatar, now massive from the inordinate number of associative bodies it keeps—845,000 ciphers, to be precise.

"You Have 0 Friends" is conversant with the seminal *Matrix* saga (1999, 2003) and its thesis re the dehumanizing effects of electronic stimulation. Directors Andrew and Lana Wachowski indulge Orwellian fantasies to disclose how people are in fact avatars controlled by a big computerized Other. In the first installment[24] hacker Neo (Keanu Reaves) swallows a red pill and becomes defiantly aware of the matrix that designs palatable realities for human pods. Thus he grapples with two Selves: his somatic body that is plugged into the machine and his experiential body that knows false diversion. The rule is, "If you're killed in the matrix, you die" in reality. Stan Marsh's hyperactive Facebook persona similarly expresses the death of his autonomy in discerning which friends appear valid and which activities, amusing. (This dilemma is equally open to parody seeing Stan's alternative to virtual networking is the insular activity of playing video games.) Science fiction narratives like these voice anxiety over the hazard of surviving outside as well as inside synthesis. Writes Derrida on hymen technology, "there is no longer any textual [or "functional"] difference between the image and the thing, the empty signifier and the full signified, the imitator and the imitated, etc."[25]

I have hinted that avatar culture may delimit everyday life with palpable consequences for virtual relationships. Besides the presence of "realness" per se, what may be lost are signals required in judging how to operate within cyberspace and by extension, the outside world. Affective cues are intensified as well as scrambled, resulting in inflations and denials of selfhood. To build on Jameson's philosophizing of political dichotomies at play in Internet use,[26] Facebook engenders a social antinomy: polygamous *and* solitary, connected *and* discrete, we fuse with many during chat sessions, fixing our bodily co-ordinates in response to a host of virtual locations. However, apart from the hypermobility of cellular technology—we can transport Facebook personas wherever our bodies go and wherever phones sustain an Internet link—tradition dictates one must be in front of the computer screen to be virtually social. For over twenty years, American cinema has implied that computer-use is fraught with the desirable insularity of the lone operator: *Revenge of the Nerds* (1984, 1987) lauds social ineptitude as it relates to mechanical genius; *Hackers* (1995) makes young and sexy the motive to find data stealthily. Finally, *The Terminator*, *RoboCop*, *The Matrix* and *I, Robot* (1984-2004) feature humans (em)powered and estranged in relation to the apparatus. With Facebook usage there is also the fantasy of being inflated then removed by technology. If online networking enables "agentic [not] communal" acts of "self-presentation," as one recent study suggests,[27] egoism is central to the mediation yet potential destruction of avatar identity. For example, to stress what they perceive as Facebook's domineering presence, Seppukoo.com facilitates "virtual identity suicide" by deactivating profiles.[28] Waggishly gesturing at the Japanese honor ritual of self-immolation, *seppuku*, Seppukoo.com provides visual drama with mock memorial pages.

More affronts register in discourses surrounding the coincident vulnerability of online and offline bodies. American news stories like the Craigslist killing (2009) and the suicide of Massachusetts teen Phoebe Prince (2010)—which set a legal precedent when Prince's peers were charged, partly, for bullying her on Facebook—publicize casualties of social media. In her cross-cultural study, *Youth Online*, Angela Thomas examines how young people consider their range of emotions, from feeling down to enthusiastic, basically personified in cyberspace.[29] The theoretical anonymity of the Internet opens avenues for non-normative social conduct along with creative, political expression, thus allowing users to take risks, hurl insults, test social taboos and rehearse real-life, often sexual scenarios.[30] In a more alarmist vein it is speculated that flirting on Facebook spurs unsafe sexual behavior: *The Daily Telegraph* reports a public health director has tied heavy usage to syphilis contagion in select British cities.[31] Thus certain genres of inquest thematize virtual communication in terms of anonymity, permissiveness, loss and excess. So computing hazards, viruses and what not, can be translated and inscribed onto the body. Indeed, human violence, abuse and infection are alleged offshoots of digital culture that warrant greater examination since they intercede in presumably disembodied milieus with sensations, physical and psychological, generative and terminal. Virtual connection, like affect, is ambivalent (recall the dual temperament of the hymen). The goal, here, is not to polarize avataristic bodies—victim vs. aggressor; neophyte vs. predator; the passive one entered vs. the active one who entered—but to question bursts from the tenuous safety of our World Wide Web, a space structurally on par with Derrida's "spider web"[32] of deconstruction, "where nothing has yet taken place" except the suspension of meaning, which implies the act that is "(at the same time or somewhere between) love and murder."[33]

How is Facebook simultaneously a domain for avatars caught up in greetings? Flirtations? Rebuffs? Following the tenet I have already stated, that avatars shelter us from and subject us to the world, one way of theorizing digital entanglements is to highlight the operator's readiness to master *autoimmunity*. Autoimmunity, in Derrida's estimation, is the ego's drive against "self-protection."[34] By obligation we welcome the unoriginal Other, or "non-ego," who contributes to our identity, making what we are mutually dependent on what we are not.[35] I wish to propose being Internet-savvy calls for autoimmunity: Facebook builds personas with an ever-growing albeit lifeless economy of friends, and, in exchange, users consent to the blurring of distinctions between private/"real" life and public/"virtual" life.

Facebook encourages an intense yet degraded sense of recognition—or rejection—whereas the film *Avatar*[36] generally sacrifices human forms for the sake of hyperreal entertainment. Evinced by its definitive title along with the "populism" of James Cameron's billion-dollar portfolio,[37] this film posits how avatars bear meaning in mass culture, specifically as motifs of capitalist desire. It is a cautionary tale of globalization gone extreme in the year 2154, depicting white American industrialists and military officials who, after pillaging their own planet, raze and romance a moon called Pandora, which is native to a blue race of humanoids called the Na'vis. These computer-generated aliens serve as attractive commodities for earthlings who crave power: through sensors in their tails they become psychically bound to their habitat's resources; besides that, they glide from trees, maneuvering their ten-foot-tall, feline-esque bodies in ways unthinkable for Jake Sully, an ex-Marine paralyzed from the waist down. Before Jake (Sam Worthington) can infiltrate Pandora's hostile atmosphere, he must enter a chamber and don an avatar built from half-human, half-humanoid DNA. While this hybrid avatar improves Jake's physical prowess by rendering him remarkably ambulatory, it also extends his role as a mere cog in military and industrial machinery. Jake trains in motions of Na'vi quotidian, learning their ways and language, therefore making it easier for corporate boss, Parker Selfridge (Giovanni Ribisi), to plot ecological genocide. *Avatar* troubles virtual connection with Marxist specters of "dead labour"[38] because Jake's artificial mimicry draws from the Na'vis' natural work in an exploitative fashion. More accurately, Jake's coveting of Pandora harnesses what Baudrillard calls "operational simulation" in that Jake cannot be a "counterfeit" version of entirely reiterable, reiterated characters.[39] Yet somehow feeling comes across, and Cameron's hero sympathizes with Pandora locals, particularly his female guide, Neytiri (voiced by Zoe Saldana), whom he finds sensual and noble. The film's only sex scene, though brief, is fecund in its virtuality: Neytiri straddles her lover while the ancestral Tree of Voices encompasses their bodies with its glowing limbs. Afterwards Jake emerges, flushed from his impasse between reality and hyperreality. "What the hell are you doing?," he asks himself. Though Jake physically and emotionally merges with Neytiri by way of his avatar, for him, their contact registers as beyond flesh, outside and contrary to instinctive ways of knowing the body. Akin to Facebook, *Avatar* deals in synthetic breakthrough *and* denial. Again, we see that technology is the site of tension where concepts of humanity, pleasure and originality converge. Though I should note Cameron attempts to bridge this chasm by generating digital figures with performance capture, a technique that employs then simulates real-life movement.[40]

Defined via the flux of invasion and stasis within hyperreality, *Avatar* is a hymeneal text on more than one level. Quite appropriately, Cameron warns us that the "web of life" is under attack.[41] Pandora is a CG screen resistant to live-action men who penetrate, and its moniker is even a feminine ethos evoking the original threat of "man's" discovery, the *Pandora's Box of hyperreality*. Subsequently Neytiri upholds the hymeneal conceit when she chooses Jake for her one true sexual partner and henceforth admits her virginity: "I am with you now, Jake. We are mated for life." By making Neytiri willing to wed a "sky person" whom she knows through prosthesis, Cameron does three things: he engineers the fierce monogamy of the Na'vis, and he ostensibly commits to the heteronormative prophylactic while aiming to rescue virtual technology from stereotypes of detachment and promiscuity (bear in mind anti-Facebook talk associating online trysts with violent acts and sexually transmitted infections). However, indicative of Derrida's and Baudrillard's belief—that simulation entails some degree of loss—the Na'vis can only be *saved*, and I use this term loosely, for much of their territory is destroyed by the denouement, if Neytiri allows Jake to be her essentially false leader against militant forces he once supported. Prior to the final showdown, she confesses, "I was afraid, Jake, for my people. I am not anymore." Her autoimmunity as a (CG) child of nature welcoming a (human) child of industry, as a not-so-real being welcoming a not-so-real avatar, is therefore manifested in all its contrariness. It is important to heed Neytiri's user status, though, to avoid altogether dismissing her phantasmic agency. She *is* accessed, like Derrida might say, in her own erasure, but she is not purely[42] an eroticized object of the hyperreal, or as the racist, sexist Colonel Quaritch dubs her, a piece of "local tail." Neytiri is also an operator who protects Jake's fledgling avatar body; who rides her sky man for sexual stimulation; and who refuses his pretense on several occasions, telling him at one turning point, "You will never be one of the people!"

Cameron's stereoscopic amalgams of digital and natural worlds strive for spatial and emotional span. *Avatar* may exhibit Brian Massumi's concept that affect is virtually "synesthetic" as it attempts to move from one given sensory mode to another.[43] Nevertheless, politically speaking, the film can be taken to task for offences against "reality." Jake's armor is his prophylactic techno-persona, which acts as a literal condom during his intercourse with Neytiri and as a symbolic shield from self-reflexive contact between imperialist and indigenous subjectivities. The Na'vis' blue skin color is the most patent sign of how *Avatar* attempts to distinguish white colonizers from those agents of color they disturb. The result is a racialization that collapses into its own defused opposition of white vs. improbable blue. What is more, to progress in Cameron's allegory of a more humane life with the humanoid, Jake simply discards his racial character and adopts his hybrid vessel, which showcases his "enlightened" Self vis-à-vis Pandora. He is evidently the white meta-avatar capable of selecting and actualizing the blue foil. Both of Jake's bodies are laid out in a genesis ritual: the final close-up shot heralds his upgrade from adjunct to essential Na'vi, and when his yellow eyes fly open, Jake becomes sentient in what he once believed was hyperreality. In this sense, Jake's reanimation is echoic of Neo's start in *The Matrix*. Through their inadvertently non-postmodern claims to alternative "truths" fully representable, both characters impart how virtual dimensions, and deluded assumptions about these dimensions revealing a more authentic reality,[44] can give way to affect. They can deliver some rupture of the real, which is the supplemental mediation of life, not wholly dissimilar from Neo's pill-induced idealogy—"A world without rules and controls, without borders or boundaries . . . where anything is possible." Nakamura notes *The Matrix* series both troubles and re-confirms white hegemony in new media, offering up Whiteness as the all-powerful yet essentially dead, over-determined interface to be resisted.[45] *Avatar* does the same when Jake relinquishes his white physicality. Yet for mainstream white audiences watching animated blue figures on a screen, the Na'vis remain the *real* avatars, even though they compose the originary material presence of Pandora, unlike the film's human "dupes."

To steer clear of disturbing a veritable rain forest in production,[46] and to lament our abusive intimacy with nature, Cameron ironically uses inorganic figures. Pandora is replete with fluorescent birds, plants and spiritual essences, which signify, to use Baudrillard's logic, how CGI is "the ultimate violence done to the image."[47] The alchemy of nature made graphic ensures *Avatar* is lucrative in a neoliberal market—Pandora is relevant to a globalized audience insofar as it supports the claim that we come to know the Other largely through idioms of technology. Scientist Grace Augustine (Sigourney Weaver) rationalizes why the sacred "home tree" cannot be bulldozed to mine unobtainium:

> I'm not talking about some kind of pagan voodoo, here. I'm talking about something REAL. Something measurable in the biology of the forest . . . What we think we know is that there is some kind of electrochemical communication between the roots of the trees.

Rather than explaining the Na'vis by way of cultural difference, an idea vaguely reduced to polytheism and sorcery in Augustine's plea, Cameron chalks their essence up to a single and quantifiable matrix of nature. Augustine continues, "It's a global network. And the Na'vi can access it. They can upload and download data, memories . . ." Augustine reveals how *Avatar*, fundamentally, confirms an era where technology is the dogma that not only legitimizes but supersedes every endemicity, every genealogical "root," regardless of how The Tree of Voices declares its otherworldly influence when Jake and Neytiri make love. Like Derrida says, "Religion today allies itself with tele-technoscience, to which it reacts with all its forces."[48] Thus I argue *Avatar*'s use of technology mitigates the green doctrine it professes to virtually materialize. The film immunizes itself against its own zeal. True, hyperreal graphics make perceptible specters of the unknown (celestial landscapes, florae and faunae). And as post-human constructs resembling biped cats crossed with superheroes, Na'vis challenge expectations for the speaking, acting being, though artificially. Still, Cameron's fetishization of avatar culture creates a paradox for his activist and Marxist auteurship,[49] as it renders the Na'vi plight liminal to immediate political sympathy when these aliens are, diegetically, viscerally, anti-technology and anti-capitalist.

But to be fair, the Na'vis are scarcely disposable. On the broadest, most uncritical symbolic plane, they stand for what people may never be in shape and what we can only hope to recover in a futuristic age of pre-industrial utopia. To be clear, I do not wish to extend the fancy that this existence was ever in our grasp. Analogous to Nakamura's idea that digital life is now more and more "diasporic"[50] and Apter's premise that avatars carry on "desubjectified agencies,"[51] the blue Na'vis are haunting figures. If Derrida will have them, they seem to be exemplary traces.[52] The Na'vis are human-made reminders of untenable human relations with the planet, the economy, teletechnology, not to mention the Self and the Other. A number of critics, unsurprisingly, insist Cameron draws out contact folklore propagated by films like *Dances with Wolves* (1990) and Disney's *Pocahontas* (1995). Debatably, *Avatar* belongs to a New World canon insisting upon the white man's dream to either violently or peacefully arrogate "foreign" perspectives by coupling with the female Other. Journalist Annalee Newitz sums it up nicely: these types of films express "a wish to lead people of color from the inside rather than from the (oppressive, white) outside."[53] Because the avatar generally possesses a hymeneal lure, it can be charged with such facility and such delusion.

This essay has interrogated how avatars embody junctions and pales of identity to espouse North America's fixation on commodifiable difference. In trying to theorize the vast possibilities of whom or what we could technically know and whom or what our Selves could technically be, users encounter worthy incongruities: motifs of blockage and passage, emptiness and excess, abjection and connection. Riddled with problems as well as pleasures, Facebook personas and CG parties of inter-species recognition give a common pledge—that, through virtuality, material agencies can tarry with inspiration as much as with death. There is room to appreciate synthesis while observing some scholarship on the violence of reproduction. Derrida's barrier politics articulate mechanical losses already present in identity; therefore, his deconstructionist work, notwithstanding its negative bent, speaks to the aim of post-reality. *To move past binaries of what is classified as real and un-real.* It is all this time spent online and onscreen with avatars musing what might be that perchance heightens and intensifies our hyperreal disparity between "human" action and "virtual" pleasure.

Notes

1. Carsten Stathausen, "The Body of the Philosopher," *New Centennial Review* 9:2 (2009): 140, 145, accessed on October 3, 2010, http://ehis.ebscohost.com/.

2. This is an interpretive application of Derridean theory.

3. Emily Apter, "Technics of the Subject: The Avatar-Drive," *Postmodern Culture* 18.2 (Jan. 2008): n. pag., accessed on May 20, 2010, http://muse.jhu.edu/.

4. Jean Baudrillard, *Symbolic Exchange and Death*, trans. Iain Hamilton Grant (Thousand Oaks, California: Sage, 1993): 71-2.

5. Ibid., 70-76. Also, see Fredric Jameson, *Postmodernism, or, The Cultural Logic of Late Capitalism* (Durham: Duke University Press, 1991), 1-54.

6. Brian Massumi, *Parables for the Virtual: Movement, Affect, Sensation* (Durham: Duke University Press, 2002), 35.

7. Jacques Derrida, "The Double Session," in *Dissemination*, trans. Barbara Johnson (Chicago: University of Chicago, 1981), 212-16.

8. I would be remiss to ignore how the hymen is traditionally understood as a heteronormative symbol. Derrida speaks of this as "the consummation of differends" [between woman and man]. See "The Double Session," 212-13.

9. One can hearken back to the early-twentieth-century work of Walter Benjamin and Siegfried Kracauer for related media critique.

10. Jean Baudrillard, *The Intelligence of Evil or the Lucidity Pact*, trans. Chris Turner (New York: Berg, 2005), 92.

11. "Statistics," Facebook Press Room, accessed February 17, 2011, http://www.facebook.com/press/info.php?statistics.

12. Lisa Nakamura, *Digitizing Race: Visual Cultures of the Internet* (Minneapolis: University of Minnesota, 2008), 17.

13. Mark Stephen Meadows, *I, Avatar: The Culture and Consequences of Having a Second Life* (Berkeley, California: New Riders, 2008), 70.

14. Here, I borrow from Judith Butler's seminal theory of performativity: avatar masquerading is based on "*fabrications* manufactured and sustained through corporeal signs and discursive means" (emphasis original). See Butler, *Gender Trouble: Feminism and The Subversion of Identity* (New York: Routledge, 1990), 136.

15. Nakamura, *Digitizing Race*, 18, 47.

16. Apter, "Technics of the Subject: The Avatar-Drive," n. pag.

17. Fredric Jameson, *Archaeologies of the Future: The Desire Called Utopia and Other Science Fictions* (New York: Verso, 2005), 164.

18. Carsten Strathausen contends the body is Derrida's main "object" of study. See "The Body of the Philosopher," 140-52.

19. Derrida states "il n'y a de hors-texte." See Derrida, *Of Grammatology*, trans. Gayatri Chakravorty Spivak (Baltimore: John Hopkins University Press, 1976), 158.

20. *South Park*, "You Have 0 Friends," MP3. Matt Stone and Trey Parker, accessed June 5, 2010, The Comedy Network.ca.

21. Because Kip is starved for any sort of affection and thus thrilled to have even one cyberfriend, I am reminded of Derrida's comments on friendship and its virtual extension: "Friendship, the being-friend—what is that, anyway? Well, it is to love *before* being loved" (emphasis original). See Derrida, *Politics of Friendship*, trans. George Collins (New York: Verso, 1997), 8.

22. Apter, "Technics of the Subject: The Avatar-Drive," n. pag.

23. *South Park*, "You Have 0 Friends," MP3.

24. *The Matrix*, DVD. Andrew and Lana Wachowski, 1999. Burbank, California: Warner Bros.

25. Derrida, "The Double Session," 209. Derrida uses terms like "functional" to dismantle binaried assumptions of difference, *not* to dismantle difference altogether.

26. Jameson, *Archaeologies of the Future*, 164.

27. Laura E. Buffardi and W. Keith Campbell, "Narcissism and Social Networking Web Sites," *Personality and Social Psychology Bulletin* 34 (2008): 1303-05, accessed on July 3, 2010, http://psp.sagepub.com/.

28. "Welcome to Seppukoo," Seppukoo, accessed on June 30, 2010, http://www.seppukoo.com. Please note that as of February 2011, Seppukoo services are no longer available "due to the paradoxical controversy between the giant Facebook and Seppukoo," Seppukoo, accessed on February 17, 2011, http://www.seppukoo.com.

29. Angela Thomas, *Youth Online: Identity and Literacy in the Digital Age* (New York: Peter Lang, 2007), 189.

30. Ibid., 184-85, 189, 177.

31. Matt Warman, "Facebook dismisses syphilis link," *The Daily Telegraph*, March 25, 2010, accessed on April 20, 2010, http://www.telegraph.co.uk/technology/facebook/7519772/Facebook-dismisses-syphilis-link.html.

32. One of Derrida's multiple names for the hymen. Derrida, "The Double Session," 224.

33. Ibid., 215, 213.

34. Derrida, "Faith and Knowledge," in *Acts of Religion*, ed. Gil Anidjar (New York: Routledge, 2002), 87.

35. Derrida, *Specters of Marx*, trans. Peggy Kamuf (New York: Routledge, 1994), 141.

36. *Avatar*, DVD. James Cameron, 2009. Century City, Los Angeles: Twentieth Century Fox.

37. James Kendrick, "Marxist Overtones in Three Films by James Cameron," *Journal of Popular Film and Television* 27.3 (Fall 1999): 37-38.

38. Karl Marx, *Capital Vol. 1*, trans. Eden and Cedar Paul (New York: E.P. Dutton & Co Inc, 1957), 232.

39. Baudrillard, *Symbolic Exchange and Death*, 56-57.

40. Rebecca Keegan notes Cameron's use of performance capture. See Keegan, *The Futurist* (New York: Crown, 2009), 236.

41. Qtd. in Ibid., 254.

42. Derrida writes, "the hymen is never pure or proper." See "The Double Session," 229.

43. Massumi, *Parables for the Virtual*, 35.

44. See Jean Baudrillard's critique of how *The Matrix* deals with the virtual and the real: "The Matrix is surely the kind of film about the matrix that the matrix would have been able to produce . . . [It] paints the picture of a monopolistic superpower . . . and then collaborates in its refraction." Baudrillard qtd. in "*The Matrix* Decoded: *Le Nouvel Observateur* Interview With Jean Baudrillard," trans. Gary Genosko and Adam Bryx, *International Journal of Baudrillard Studies* 1.2 (July 2004), n. pag., accessed on June 5, 2010, http://www.ubishops.ca/baudrillardstudies/vol1_2/index.html.

45. Nakamura, *Digitizing Race*, 98.

46. Elizabeth Renzetti, "James Cameron's Avatar: A symphony in blue and green," *Globe and Mail: Arts*, last updated January 12, 2010, accessed July 10, 2010,

http://www.theglobeandmail.com/news/arts/james-camerons-avatar-a-symphony-in-blue-and-green/article1405271/email/.

47. Baudrillard, *The Intelligence of Evil or the Lucidity Pact*, 95.

48. Derrida, "Faith and Knowledge," 82.

49. See James Kendrick's Marxist reading of *Titanic, The Abyss* and *Aliens* in "Marxist Overtones," 36-44.

50. Nakamura, *Digitizing Race*, 47.

51. Apter, "Technics of the Subject: The Avatar-Drive," n. pag.

52. See Derrida's work on *la trâce*: "as soon as there is a trace, there is also some virtualization; these are the abc's of deconstruction." Derrida, *Without Alibi*, ed. and trans. Peggy Kamuf (Stanford: Stanford University Press, 2002), 210.

53. Annalee Newitz, "When Will White People Stop Making Movies Like 'Avatar'?," *i09*, December 2009, accessed on August 20, 2010, http://io9.com/5422666/when-will-white-people-stop-making-movies-like-avatar.

Part III

The Game of Life: Playing in the Hyperreal

Chapter 9

Planting Crops in the Hyperreal: *Farmville* and Simulated Work

Emily Hall

For many, simulations can be seductive, as they offer everyone the ability to assume a new identity and to escape a mundane life. The danger with such a seductive appeal is that, at times, those involved in simulations begin to view the simulation as the "real world," and in turn, replace the real with the hyperreal. Such was the case in early 2010, when a Bulgarian government official was ousted from his position after he abandoned his work tasks in order to cultivate his virtual farm in the internet game *Farmville*. On March 31, 2010, National Public Radio (NPR) ran a radio broadcast in which it was announced that several Bulgarian officials "were caught playing [Farmville] during meetings. But only one lost his job. The council decided he had spent too much time tending his virtual crops and not enough time tending to city affairs."[1] If the Bulgarian official's dedication to his virtual farm was an isolated event, then his devotion to *Farmville* could be dismissed as a singular occurrence due to a lackadaisical work ethic. However, in a BBC article entitled "Down on the Farmville," Rajini Vaidyanathan states that in less than one year, *Farmville* has managed to accumulate an astounding "80 million people."[2] Undoubtedly, the popularity of *Farmville* stems from the fact that game play is free. Yet the surge in *Farmville* players, compounded with the firing of a government official, makes *Farmville* ripe for a cultural studies analysis. It is obvious when one begins to analyze the *Farmville* phenomenon that the creators of the virtual game have constructed a hyperreality. As the game offers a myriad of awards, prizes, and sometimes even notable recognition, players can come to believe that they are more appreciated at their simulated job than at their real job. Thus, the hyperreality of work created in *Farmville* supersedes work in the real world.

Constructing a hyperreality is no facile task, as the goal of a simulated world is to be as reflective of the real world as possible. Umberto Eco in *Travels in Hyperreality* notes that in order to create an effective hyperreality, those that are fabricating it must construct a mimetic representation on a vastly smaller scale.[3] Additionally, he asserts that "if a reconstruction is to be credible, it must be absolutely iconic, a perfect likeness, a 'real' copy of the reality being represented."[4] In *Farmville*, the creators of the game attempt to simulate agricultural work. The player begins the game with a tiny farm, a small amount of coins, and a handful of seeds. The point of the game is to amass both great wealth and as many neighbors as possible. Players may also fill their farm with a myriad of animals, which each provide the farmer with a unique crop, and trees, which yield fruit. While harvesting trees and animals do not result in a boost of experience points, they do add to the farmer's coffers. Players gain experience points by plowing and planting crops, and after the points have accumulated to a specified amount, the players advance another level. With each new level typically comes the ability to plant a new, much more profitable crop, and the chance to unlock several items with which the gamer can decorate his farm. As the game is set to real-world time, crops wilt and die if players are not dedicated enough to return to the game every few hours. The game does not have a culminating objective, and the player is merely invited to play the game at will and advance at leisure.

When the gamer first accesses the game, he is invited to create an avatar of himself, which will appear on the screen and perform all the agricultural tasks. There is not simply a female or male avatar option;[5] instead, the player can customize his avatar by selecting the avatar's hair style, hair color, facial features, eye color, and skin color. The idea is that the player may either select a character that is visually reminiscent of his own features, or he may adopt an avatar that has physical traits he has always desired. Thus, the gamer has the choice: does his *Farmville* avatar reflect his real appearance or does he craft a new appearance in the virtual world?

Some may argue that if the player selects an avatar that is mimetic of his true appearance, he is staying anchored in the "real" world. After all, he is not assuming a new exterior or identity, but rather, making the conscious decision to represent his actual self. To extend this syllogism further, in the antithetical scenario, he is assuming a false identity and is therefore slipping into the hyperreal. Yet Slavoj Zizek in *The Plague of Fantasies* argues that in both scenarios, the participant in the virtual reality setting still falls prey to the delusions of the hyperreal. For example, while it would appear that the gamer that chooses an identical avatar is making a harmless selection, Zizek propounds that he still compromises his sense of reality. While Zizek notes, "On the one hand, we maintain an attitude of external distance, of playing with false images" he goes on to state that the player then places this identity into the realm of the hyperreality; he, in other words, places a *copy* of himself into the virtual simulation. Meanwhile, Zizek fervently believes that when a player purposefully chooses a façade that is radically different than his actual self, he exhibits an intent to wholly lose himself within the simulation: "On the other hand, the screen persona I create for myself can be 'more myself than my real-life' persona."[6] In this scenario, the gamer utilizes the avatar to project his ideal self, his preferred physical appearance, and controls how he would wish to be perceived. Regardless of whether or not the player selects a mimetic avatar, he begins to identify with his character on screen. As Zizek observes, the player becomes tempted to think, "it's nice from time to time, to forget one's true self and put on a more satisfying mask—this way you are delivered of the burden of being what you are, of living with yourself, and being fully responsible with it."[7] The avatar, then, becomes a vehicle through which the player may escape the pressure of the real life work force and enjoy the simpler demands of online farming.

Once the gamer establishes his avatar's physical features, he must begin to "farm." While he may level up by growing crops, he cannot truly achieve agricultural success until he gains numerous neighbors. In *Farmville*, players expand the number of neighbors by inviting their Facebook friends to join the game. Each player sends an online invitation to his Facebook friends to join the relaxing world of online farming. The idea is to build a community of players in which the *Farmville* farmer is surrounded by his friends as he plants his crops. At the bottom of the screen the neighbors are ranked in ascending order according to levels. The Facebook profile picture and the picture of the avatar of each player are displayed and the option is given to visit the neighbor's farm. When the game first appeared in 2009, the only way that the *Farmville* farmer interacted with his neighbors would be to visit the neighbors' farms and to tend to them, which allowed the farmer to gain a few experience points. A few months after the game began to attract popularity, more features relating to the neighbors began to surface. For example, each player was given a free chicken coop by the creators of *Farmville.* The idea was to place the chicken coop on the farm and to predictably fill the coop with chickens, which are the cheapest animal to purchase in *Farmville.* Then, after the farmer fills his coop with his fowl, his neighbors would kindly feed the chickens. Naturally, a reward is provided for feeding the chickens, and neighbors could win anything from a chicken to place in their own coop to a sought-after decoration like a merry-go-round, which could be purchased only with *Farmville* cash.

In April, 2010 neighbors became incredibly important and highly lucrative. *Farmville* farmers were provided the option to join co-ops with their neighbors. For the co-op an actual target goal was given and neighbors were urged to band together in order to fulfill the requirements of the objective. Typically, in each co-op scenario, a time limit of so many hours/days was given, and within the time span, the collection of neighbors had to plant and harvest a specific amount of one (or more) crop(s). As usual, a reward was given if the neighbors were able to complete the task before the time limit. The reward, similar to the task of feeding chickens, was a decoration that was inaccessible without using *Farmville* cash.

Yet deadlines, quotas, and fellow employees *seem* like work. Indeed, it is work, as the *Farmville* farmer had to ensure that his crops were harvested within the exact time span, or else his co-op assumed a collective loss. Thus, it seems strange that the player would indulge in a co-op setting, which has the ironic possibility to create a stressful game playing environment. However, several factors collide within the co-op scenario. One reason as to why co-ops in the simulation game could appear more palatable than real work is that the gamer could select which co-op he could join, as several scenarios and possible time limits were presented. Second, the *Farmville* farmer is under the illusion that he works with his "friends" or neighbors. In a real-world employment situation, the employee rarely has any input as to those he will work with. Within the perimeters of the *Farmville* co-op, he is able to specifically select cohorts from his list of neighbors to be his fellow farmers. In *Simulacra and Simulation*, Jean Baudrillard posits that simulations often cause human relations to be reduced to the level of the hyperreal: "The social itself is residual once it becomes a production of 'social relations.'"[8] In other words, once a company manipulates social relations for profit, then society is at the whim of the hyperreal. In *Farmville*, the creators construct a simulation in which one must assemble a team of neighbors in order to work towards a common goal. As a result, human relations are reproduced on a small scale, as the avatars "help" one another in order to ease the burden of the work.

It is important to note that *Farmville* cannot nurture real human relations. Other than the ability to join co-ops, the only other way that one may keep in contact with one's neighbors is through the ability to leave messages for them to view. When the game first premiered, neighbors could leave tiny sign posts with a short message on each others' farms. The posts would be planted somewhere on the neighbor's farm so that he could find them and read the messages. At the beginning of 2010, the messaging system was amplified. Whenever the gamer would go to play *Farmville*, an entire message board would appear on the screen that would list off the numerous messages left by his neighbors. The messages may be personal messages from neighbors who leave a specific note for the intended person, or a standard, *Farmville*-generated message that states which neighbor helped on the farm. In this sense, the creators of *Farmville* attempt to make an interconnected web of players, an entire community of *Farmville* neighbors who are all "working" together. Thus, the gamer comes to believe that his fellow neighbors are helping him work harder on his farm, either though encouraging messages that are left via the message board or through aiding him with menial tasks.

However, such lines of thinking are erroneous, as the gamer is invited to suspend reality and believe that camaraderie and human relations can be reduced to online co-ops and message boards. Zizek propounds that the perpetuation of cyber-simulation relationships are a result of people willing to enter into shallow relationships in order to lose themselves in the hyperreality: "The problem with today's social functioning of cyberspace is thus: that it potentially fills in the gap between the subject's public [. . .] identity and its phantasmic background."[9] As Zizek suggests, real human relations cannot be fostered in simulations, as the relations are channeled through avatars or simulated representations of the person's identity. The person who plays *Farmville* is *not* the avatar on the screen, nor do the avatars of the neighbors properly embody the neighbors. Thus, as the avatars represent copies of the individuals, human relations become mere imitations as well, and the "neighbors" and "co-op coworkers" in *Farmville* are mere hyperrealistic affairs.

The *Farmville* gamer is lured into working on his simulated farm in order to indulge his desire to create a new identity and to hand-pick "neighbors" on whom he may rely to help him work on his farm. While the chance to construct an alternate identity and to accrue numerous "friendships" could be enough to ensnare a gamer into simulated work environment, in order to properly trap the gamer in the simulation, he must continuously return to the game. In *Farmville*, numerous awards and prizes are bestowed upon dedicated workers. The most obvious motivation in the game is to increase experience points to obtain the next level. As in a real work setting, putting effort into the simulated farming will allow the gamer to quickly rise in rank. In *Farmville*, every time the next level is achieved it can be posted onto Facebook to allow friends on the social networking site to know of the accomplishment and to lavish praise on the recipient. Additionally, *Farmville* offers the gamer the ability to earn ribbons. Typically they are earned through accumulating so much of an item, crop, or animal. Similar to the manner in which the gamer may announce that he is on a new level, *Farmville* provides the option to announce whenever a new ribbon has been awarded. Yet like all things in *Farmville*, the ribbons are only simulations of real accolades. Not only does the ribbon itself only exist in the cyber-simulation, but the ribbon is awarded for *simulated* work. Thus, the *Farmville* farmer's work is rewarded with imitations of actual awards, which require little to no effort to achieve. Eco avers that as long as the copy is reminiscent of the actual, mankind will be "taken in by the verisimilitude" and continuously return to the hyperreal[10]. In *Farmville*, just as in real life, incentives are offered to the worker, but true to the tenets of hyperreality, these incentives are cheap reproductions of the real, meant only to whet the appetite of the player and to make the copy seem as life-like as the actual.

Along with amassing work-related accolades and moving up in rank, *Farmville* offers money-related incentives for those willing to work for them. Whenever the gamer harvests his crops, he is rewarded with a cash prize. In the beginning of the game, the types of crops that he is able to grow (which are directly related to how far he has advanced in the game) yield him only a paltry sum of coins. Thus, in order to earn a significant amount of money, the player must vigilantly plant and harvest crops for numerous consecutive weeks. Similar to real life, if the *Farmville* farmer is unwilling to work on his farm, he is not given any money and has no way to support himself in the game. Currency in *Farmville* comes in two forms: cash and coins. Besides the occasional gift from the *Farmville* creators, *Farmville* cash must be purchased with actual credit cards and with real money.[11]The coins are distributed to the farmer immediately after crops are reaped. The gamer then has a plethora of ways to spend this money. He may choose to purchase only his seeds, farm animals, and necessary agricultural machinery, such as harvesters and biplanes. The much more appealing option is to spend the coins on the myriad of decorations that he may place on his farm.

The decorations that the gamer can purchase extend beyond a nice fence or a home. Indeed, the *Farmville* farmer is urged to landscape his entire farm and to make it seem as much of a resort as possible. Seasonal themes are provided so the farmer can select from arctic themes in winter, romantic themes in February, and sprightly Easter themes around the beginning of spring. The themes are so expansive and inclusive that the gamer can design his farm to be an exact replica of ancient China or Italy. The possible themes for June, 2010 were Tuscany, Italy and a tropical theme. For those interested in the Tuscan theme, they could purchase decorations, buildings, and crops that hailed from that region. If the *Farmville* cultivator wanted to make his farm a tropical destination, he could festoon his farm with, amongst many things, huts, pigs in hula skirts, surf shacks, and even a blanket of "sand" that would replace the ordinary green "grass" that stretches across his farm. The result of such intricate detailing is that *Farmville* becomes less about idle farming, and more about creating a world that is meant to be an escape from reality. Eco notes that a hyperreality essentially "creates a network of references and influences that finally spread also to the products of high culture."[12] By allowing *Farmville* users to purchase famous architectural structures from other countries and culturally significant items, the creators are utilizing the "references" in order to make the players feel as though they are creating the locations on a vastly smaller scale. Thus, while the worker cannot actually visit the real Tuscan region of Italy, he can construct a miniature Tuscany on his farm. The addictive appeal of hyperrealities, then, is a result of what Eco believes is the hyperreality's ability to replace the real: "we are giving you the reproduction of [it] so you will no longer feel any need for the original."[13] In other words, *Farmville* offers the chance to "work" hard and earn a getaway to a foreign land, and while the destination resort is merely on a simulated farm, if one can purchase enough representative pieces, one's simulated world will become a more convincing copy of the originals found in the real world.

As previously stated, the *Farmville* cultivator has the option to purchase *Farmville* cash that can be used to buy items for his farm. Typically if the gamer desires to complete an entire theme, he has no other option than to use his real money to buy *Farmville* currency. Anyone who willingly exchanges his actual money for cash that only has value in a simulation completely immerses himself within the hyperreal.[14] Thus, the danger of hyperreality, as Baudrillard observes is when "simulation threatens the difference between the 'true' and the 'false,' the 'real' and the 'imaginary.'"[15] The creators of *Farmville* fabricated a simulation game that lures in players so that they may, in turn, profit financially. The more a player is pulled into the world of *Farmville*, the more likely he is to buy numerous items for his farm with *Farmville* cash. Indeed, the creators suggest to the gamer that if he merely works diligently enough, his farm will be a respite from the real world. Eco proposes that if a simulation can appear "full" and is sufficiently representative of the real world, then people will always select the hyperreal over the real. Additionally, he notes that a hyperreality is formed whenever a company "sells copies."[16] In *Farmville*, several copies are sold by the creators, as currency is copied, cultural icons are replicated and agricultural work is simulated. Naturally, the recipients of the benefits of the work that players perform are the creators of *Farmville* themselves, who have manufactured such a life-like game to increase the odds that the gamers are willing to spend real money on it.

As *Farmville* offers an escape in return for a miniscule amount of work, it is little wonder that the Bulgarian official chose to ignore his real work tasks in lieu of laboring on his simulated farm. However, perils arise when people decide to abandon the real in order to live solely in the simulation. When people begin to devote vast segments of time to simulations, it is only natural that the real will appear less attractive: "reality will always seem inferior to it."[17] Work in *Farmville* proffers many more advantages to work in the actual world, and this, compounded with the ability to assume a new identity within the game, can entice workers from their real careers. For most, working at an actual job will not result in earning a plethora of awards or recognition. Additionally, the likelihood that working in the real world will allow a person to purchase an idealized getaway is miniscule. *Farmville* offers its players the chance to escape their ordinary lives and to live and work in a simulated realm in which performing simple agricultural chores results in great wealth and acknowledgment through rewards. In return, it asks that players suspend reality and give up their real time and money to live in a hyperreality, where happiness can be unearthed by saving up to buy a new simulated plow.

Notes

1. "Bulgarian Official Fired for Playing Farmville." Morning Edition. National Public Radio (31 March 2010) Radio.

2. Vaidyanathan, Rajini, "Down on the Farmville," *BBC News Magazine,* March 25, 2010, accessed on June 19, 2011 http://news.bbc.co.uk/2/hi/uk_news/magazine/-8585999.stm.

3. Umberto Eco, *Travels in Hyperreality*, Trans. William Weaver (San Diego: Harcourt, 1986), 4.

4. Eco, *Travels*, 4.

5. In certain agricultural simulations games, such as Harvest Moon, players may only choose from a boy or girl avatar with pre-selected features.

6. Slavoj Zizek, *The Plague of Fantasies*, (London: Verso: 2008), 176.

7. Zizek, *Plague*, 176.

8. Jean Baudrillard, *Simulacra and Simulation*, Trans. Sheila Faria Glaser (Ann Arbor: University of Michigan Press, 2010), 146.

9. Zizek, *Plague*, 212.

10. Eco, *Travels,* 45.

11. The *Farmville* "cash" is not a direct one-for-one ratio, but rather, fifty American dollars can buy as much as three-hundred *Farmville* dollars, and is advertised on the website as a "deal." By advertising the ratio as a deal, players are assured that *Farmville* cash is just as real as actual money. Additionally, *Farmville* gift cards can be purchased in a variety of locations, which can be "cashed in" online for *Farmville* cash.

12. Eco, *Travels*, 4.

13. Eco, *Travels*, 9.

14. Interestingly, while players can transform their real currency into simulated *Farmville* currency, the symbolic value of currency cannot be simulated. In other words, real money is merely paper that has an accepted symbolic value. As *Farmville* is a simulation, the currency is intangible and lacks symbolic value. Thus, once again, the hyperreal does not have what Zizek calls "depth and value" (170).

15. Baudrillard, *Simulacra*, 13.

16. Eco, *Travels*, 43.

17. Ibid., 43.

Chapter 10
Fear and Loathing in *Second Life*: Body Surveillance in the Online Community

Cathie LeBlanc

In the early 2000s, the online community *Second Life*[1] captured the popular imagination through its use of avatars, cartoon representations of the inhabitants of the community. Numerous news stories[2,3,4] touted the online community as a place where avatars would remove the limitations of real bodies from human communication. No longer would we suffer from discrimination based on gender, race, age, body type and so on because bodies in the online community are not "real." In a job interview, for example, the interviewer can make no assumptions about "real" race from the race of the avatar of the interviewee. Far from being a utopia where real bodies do not matter, however, I will argue that *Second Life* makes explicit the self-surveillance—viewing the body as an object with deficiencies, each of which can and must be fixed—that has become an implicit part of modern life. Using Foucauldian notions of power and observation as well as Althusser's notion of ideological state apparatuses, I will explore the implicit self-surveillance encouraged by popular makeover shows such as *Extreme Makeover*. I will then show that this implicit self-surveillance is made explicit in the creation and customization of avatars in *Second Life*.

Second Life was launched June 23, 2003.[5] As the site become increasingly popular, news stories touted the site as a mechanism for bringing folks together for job interviews and shared cultural experiences in cyberspace. This online community and its possibilities became a plot point on a number of television shows, including *The Office*, *CSI New York* and *Law and Order: Special Victims' Unit*. In 2007, the number of avatars created in *Second Life* surpassed seven million.[6] The owners of those avatars, called *residents*, have a multitude of options as they create and customize the body that will represent them within the community.

Just as *Second Life* was emerging, a seemingly unrelated phenomenon was occurring. In 2002, the reality TV show *Extreme Makeover* premiered on ABC, ultimately airing fifty-five episodes.[7] Personal makeovers have long been a mainstay of daytime television talk shows but the introduction of *Extreme Makeover* to the primetime lineup brought the makeover to the mainstream and pushed the limits of the genre. Body modification via cosmetic surgery had not typically been a part of the traditional makeover but such surgery became an integral part of the new makeover shows such as *Extreme Makeover* and its rival, *The Swan*. The premise of these makeovers is that external appearances represent internal realities. In other words, happy people of good character look good on the outside. This idea plays out in one of two ways on these shows. Either a makeover recipient can change her appearance in order to change her character, to make herself happier, or she can change her external appearance so that it matches her internal beauty.[8] In both cases, the external must "match" the internal and it is the external that must be changed.

A typical episode of *Extreme Makeover* chronicles the stories of two makeover recipients, almost always women. These women are deemed worthy of a makeover by virtue of the hardships that life has forced them to face. Sandra's story, which aired in the show's first season, is typical.[9] Sandra spent many years caring for her husband who had been disabled while acting in the line of duty as a police officer. In this episode, he has recently died and now Sandra is able to focus on herself, able to transform her outward appearance to match her internal beauty. We watch as Sandra meets with Dr. Ron Perlman, her cosmetic surgeon. Perlman makes his observations of Sandra's face and body visible, for the benefit of the viewing audience as well as for Sandra's benefit, via the use of a black marker. He draws lines on her face, indicating the places where he will cut into her flesh and fix what he deems to be deficient. Her body becomes an object to be examined with the expertise of a cosmetic surgeon who will then use his (almost always "his") expertise to fix the deficiencies he observes.

Extreme Makeover is emphasizing what Michel Foucault called the *medical gaze*. In *The Birth of the Clinic,*[10] Foucault discusses the idea that doctors are given special privilege in our society. That is, we have come to believe that they are able, by virtue of practice, to see, simply via observation, what is wrong with patients. They gaze at patients and observe the disease that is part of us all. This medical gaze puts a body under surveillance as an object to be diagnosed and, ultimately, fixed. Human bodies are seen in terms of their parts, that is, as an object made up of other, smaller objects, each of which has the potential (and indeed, the likelihood) of becoming diseased or deficient in some way. When Dr. Perlman marks Sandra's face and chest, he is making his medical gaze visible. We, the viewing audience, become complicit in this objectification as we observe and are asked to acknowledge the deficiencies in this body before us.

Our point of view in this process is that of the doctor, with the special advantage of the medical gaze.

Before Sandra ever arrives at Perlman's office, however, we watch as she identifies some of the deficiencies on her face and body that she would like to fix. She, like us, has in other situations, gazed upon bodies from the doctor's point of view, with the objectifying medical gaze. Because of this previous practice, she has begun to internalize the medical gaze, no longer requiring the doctor to find flaws that must be addressed. In other words, she has begun to engage in self-surveillance and self-objectification. And we, the viewing audience, are not immune to the internalization of the medical gaze. We all begin to be critical of our bodies and faces, finding deficiencies and flaws that we would like to fix.

Second Life provides an environment where an individual's self-surveillance is made visible and explicit, just as the surgeon's black marker makes the medical gaze visible and explicit. Each *Second Life* resident is required to create an avatar, a cartoon representation of her body that will interact with the objects, spaces and other avatars that make up the physical (online) environment of the community. The resident is presented with a variety of default bodies from which to choose but learns very quickly that the default bodies are not acceptable within the community because they mark the resident as a *newbie*, a new user who does not yet quite belong. The resident, therefore, is encouraged (by the disciplinary power of the community, which will be discussed later in this essay) to customize the default avatar to suit her real character. In other words, she must make the external match the internal by changing the external.

Within *Second Life*, every part of the body of the avatar can be customized. When the resident chooses to edit her appearance, the body of her avatar becomes inert, facing outward from the screen, looking toward the person it represents. The *Second Life* resident sits outside the screen, observing this cartoon representation of herself. The resident sees the avatar's body as an object separate from herself and yet this object *is* her within the online community. This objectification of the avatar is further enhanced via the methods used by the resident to modify and customize the avatar.

Modification and customization of the avatar is encouraged via a simple set of tools in *Second Life*. A series of slider bars is presented next to the inert avatar, one for each attribute of the body. Using these slider bars, the resident can change the color of the skin, the amount of freckles on the skin, the size of the breasts, the length of the arms and legs, the size and shape of the mouth, the color, size and shape of the eyes, the size and shape of the nose, the thickness of the waist, the broadness of the shoulders and so on. There is an infinite variety of combinations of attributes which encourages tweaking each attribute until it is just right. The resident here performs a function similar to the plastic surgeon in *Extreme Makeover*. She identifies parts of the body that are deficient, that don't match the image of herself that she carries in her head, and she makes corrections. Unlike the surgeon, she does not need to mark the imperfect body part first. She simply makes the change, over and over, until it is just right. In *Second Life*, no plastic surgeon is needed to place residents under the medical gaze. Each resident can be her own plastic surgeon, in a kind of do-it-yourself ethic of fixing her own body. The medical gaze is part of the ideology of the online community. Each resident assumes, without question, that she will observe her avatar and fix the deficiencies (for there will be deficiencies) in the body.

The resident's avatar does not have to be inert for the resident to observe herself as an object separate from herself. Even while the resident is navigating through the online community, she is encouraged toward self-surveillance. As the avatar moves from location to location within *Second Life*, the point of view on the screen, as presented to the resident, is from just behind the avatar's head. The resident sees what is in front of the avatar, in the direction that the avatar is looking, but the resident also sees the back of the avatar's head. The resident is always observing herself (her avatar) from behind. The resident's representation is again both her and separate from her. The effect is a feeling of always being under surveillance, even if that surveillance is self-surveillance. No matter where the avatar goes, she is always observed, from behind, by the resident. One cannot help but feel that the avatar is always under surveillance because the avatar *is* always under surveillance—by the resident herself.

In *Discipline and Punish,*[11] Michel Foucault explains the process that leads us to feel that we are always under surveillance, which causes us to behave as though we really are always under surveillance, even when we are not. Foucault explains that, at the time of the French Revolution, the traditional method of punishment meted out by a sovereign power, a specific authority which disciplines its subjects visibly and with force, was losing its effectiveness. Foucault then describes *disciplinary power*, a power that takes the place of sovereign power. The development of the panopticon, a kind of prison in which the jailor can see each prisoner without being seen himself, results in prisoners who behave as though they are always under surveillance, even when they are not actually under surveillance. Because the prisoners then behave as though they are under surveillance, there is no longer a need to keep them under surveillance at all times. When the surveillance feels as though it is pervasive, the behaviors desired by those in power do not have to be coerced. Instead, each individual internalizes the desired behavior, consenting to behave "appropriately." The prison is therefore a much easier place to run.[12]

Second Life reinforces the feeling that there is a disciplinary power, because in this online community, the individual always watches her avatar move through the world, creating a sense of the body as both subject and object. The resulting objectification of the online body encourages further self-surveillance to ensure that the body conforms to the norms of the online community.

Althusser furthers Foucault's explanation of coercion by explaining how it works outside the prison environment, in the larger society.[13] Some state apparatuses, Althusser tells us, such as prisons, the legal system and the political system, rule by violence. These are *repressive* state apparatuses, which primarily employ repression (similar to Foucault's idea of sovereign power) and secondarily use ideology (similar to Foucault's idea of disciplinary power) to maintain the power of the state. Althusser distinguishes repressive state apparatuses from *ideological* state apparatuses, such as churches, the educational system and the media. Ideological state apparatuses primarily employ ideology and secondarily use repression to maintain the power of the state. Because the ruling class is represented by the state, it is the interests of the ruling class that are maintained by these state apparatuses, especially the ideological ones.

The difficulty of ideological state apparatuses is that they function largely invisibly in the larger society, making us think that the implicit values and rules in this ideology are not actually ideology but are instead simply "the way the world works." The ideological state apparatuses rule by consent. That is, individuals within the society accept the ruling ideology as a given. They give implicit consent to being ruled by this ideology. An example from *Extreme Makeover* illustrates this point. In the episode previously discussed ("Karen and Sandra"), Sandra, who has been caring for her disabled husband until he recently died, confers with Dr. Perlman, our benevolent plastic surgeon. Her face is marked with ink to indicate the flaws, the things that need to be fixed. Dr. Perlman discusses her upcoming breast enhancement surgery, casually touching her breasts as though they were inanimate objects, part of the inanimate object that is her body. His medical training allows him this privilege. But then he delivers the bad news. She will not be able to undergo the breast enhancement surgery until the raging infections and abscesses in her mouth, teeth and gums are dealt with. Sandra is understandably disappointed at this delay but Dr. Perlman assures her that her dental issues will be corrected and before she knows it, she will have her breasts enhanced.

When I discuss this particular portion of the episode with my students, I ask them why they think that Sandra has such significant dental health issues. Without fail, they answer that it is because she has poor hygiene, that the situation in which she finds herself is the result of some sort of personal failure. They do not question the fact that we live in a society in which regular dental health care is a luxury, something that Sandra could not afford. The ideology of our society is largely invisible to them because they see this as simply the way the world works. They do not question why a makeover show would be so focused on the exterior appearance of this person without attempting to address (or really even discussing) the economic difficulties she faces. The show makes no attempt to provide her with an education or a new job or regular health care because raising those issues would make the economic inequities in our society visible in a way that runs counter to the interests of the ruling class. When the ruling power's ideology is pervasive in this way, Althusser tells us that it becomes internalized in each individual within the society as an understanding of the way the world works.[14] And this internalization of that ideology results in a lack of questioning about the way the world works, which is in the interest of the ruling class.

Second Life makes explicit many of these implicit conditions of modern life. The way *Second Life* works is accepted as the way any online world *must* work. As a resident enters this world, for example, there is no questioning of the idea that the avatar must be customized. The notion that being a newbie in an online world is a bad thing, a thing to be avoided, is enforced immediately upon entering the world when it becomes clear that knowledge of the world and its workings means power within the world. New residents, therefore, work to achieve as much knowledge (or at least the appearance of having such knowledge) as possible about the world. The quickest way to be identified as a newbie is to wander through the online landscape with an avatar body that is very close to one of the default bodies. The culture nearly demands that every avatar body be customized, since one's external appearance (society tells us) is so closely linked to one's internal state. One cannot be knowledgeable about *Second Life* if one does not *appear* to be knowledgeable about *Second Life*. One cannot appear to be knowledgeable about *Second Life* in a newbie (uncustomized) body. Customization requires that the resident examine her avatar body with the gaze of a plastic surgeon, an exacting medical gaze, to fix what is wrong. In addition, there is no questioning of the resident's point of view as she wanders around. Because the avatar *is* her, she has a first person point of view. But this first person point of view allows her to see the back of her own head, so this body that is her is also not her. The body is external to her and as a resident of *Second Life*, she accepts without question that she will be able to see the back of her avatar's head as she moves around the world. These unquestioned assumptions about the way *Second Life* works, about the culture of the online community, represent an ideology that values external appearances, that sees those external appearances as both a part of and separate from the internal. The resulting objectification of the body in *Second Life* both mirrors and enforces, via Foucauldian disciplinary power, the objectification of bodies in the real world.

Second Life is touted as a world in which real bodies no longer matter. Its proponents tout it as a space in which race, gender, age and physical impairments can be ignored. A resident creates an avatar to represent herself in interactions with other residents of the online community without the limitations of real bodies. As a cultural artifact, however, the online community cannot leave the real behind. *Second Life*'s creators impose an ideology on its residents, an ideology which cannot be separated from ideologies in the real world. As a result, *Second Life* both mirrors and enforces the objectification of bodies that is encouraged in the real world. We spend our time observing, examining, customizing, fixing our external appearances rather than focusing on inequities and power differentials within society. *Second Life*, as a product of Western consumerist culture, does not actually free us from that culture. By encouraging us to accept the ideology of the online community as "the way the world works," *Second Life* encourages us to accept the ideology of Western consumerist culture as "the way the world works."

Notes

1. *Second Life*, accessed June 17, 2010, http://secondlife.com/.

2. Mark Baard, "Virtual Business, for Real? It's Cheaper to Meet in Second Life Than in a Ballroom," *Boston Globe*, November 10, 2008, B5.

3. Benjamin Duranske, "The 3d Internet Will Change How We Live," *Wall Street Journal*, July 17, 2008, Eastern ed., A13.

4. Tom A. Peter, "Study Abroad through Second Life," *Christian Science Monitor*, July 3, 2008, 25.

5. *Wikipedia: Second Life*, accessed June 10, 2010, http://en.wikipedia.org/wiki/Second_life.

6. Frank Rose, "Lonely Planet," *Wired*, August 2007, 141-44.

7. *Wikipedia: Extreme Makeover*, accessed June 10, 2010, http://en.wikipedia.org/wiki/Extreme_Makeover.

8. Cressida J. Heyes, "Cosmetic Surgery and the Televisual Makeover: A Foucauldian Feminist Reading," *Feminist Media Studies* 7.1 (2007): 17-32.

9. "Karen and Sandra," *Extreme Makeover*, United States: ABC, 2003.

10. Michel Foucault, *The Birth of the Clinic: An Archaeology of Medical Perception*, trans. A.M. Sheridan Smith (New York: Pantheon, 1973).

11. Michel Foucault, *Discipline and Punish: The Birth of the Prison*, trans. Alan Sheridan (New York: Pantheon, 1977).

12. Foucault, *Discipline and Punish: The Birth of the Prison*.

13. Louis Althusser, "Ideology and Ideological State Apparatuses (Notes Towards an Investigation)" (1970), in *Lenin and Philosophy and other Essays*, trans. Ben Brewster (1971), 121-176, accessed June 19, 2010, http://www.marxists.org/reference/archive/althusser/1970/ideology.htm.

14. Althusser, "Ideology and Ideological State Apparatuses (Notes Towards an Investigation)."

Chapter 11

Total Immersion and the Total Screen: The Simulated and Screened Realities of Video Games

Katheryn Wright

From the chaotic image of static snow on your television, an eye emerges. As the virtual camera pulls back to widen the view, this imagery starts to make sense. The eye belongs to a woman, quite literally, on display. She is a moving image on a transparent two-dimensional holographic screen. The onscreen space continues to expand as the virtual camera zooms out further to reveal a group of passengers on a ship, the USG Kellion, watching a pre-recorded distress signal from Nicole Brennan, a crewmember of the USG Ishimura. The plot of the third-person shooter game *Dead Space* revolves around a rescue attempt by Nicole's boyfriend, Isaac Clarke, the main character who must fight the zombie-like creatures called Necromorphs. When the Kellion crashes onto the Ishimura, another screen appears within the windshield of the rescue vessel. As Isaac Clarke (the perspective the player assumes) stands after the crash, the actual gameplay begins and the player discovers that holographic projections enabled by his special suit replace the more conventional head-up display (HUD) used in a majority of third-person shooters.

The opening sequence of *Dead Space* introduces the virtual world of the video game through the layering of several screens. First, the player's television screen frames the static snow. Second, the crew uses a futuristic display device to watch the distress signal. Third, the windshield turned interface conveys technical information about the ship. And fourth, the player utilizes the holographic projection capabilities of Isaac's suit to navigate through the game. The central role the screen plays in this opening sequence, and its continued importance throughout the course of the video game, illustrates how the simulated reality of *Dead Space* is simultaneously a screened one. Working alongside Jean Baudrillard's concept of the total screen, this chapter analyzes the simulated and screened realities of several narrative games like *Dead Space* where the player uses the screen (or series of screens) to maneuver within simulated environments.

The concept of simulation, especially in the context of gaming, implies two related yet distinct ideas. Stemming from the work of Jean Baudrillard, the theory of simulation focuses on the replacement of reality with a simulated or hyperreality. Often conceived of as merely a copy or imitation, simulations are no less real. The difference is that the materialism of simulated reality is rooted in a play of signs and codes—of simulacra—rather than the material realities of labor. Screens play a key role in Baudrillard's theory of simulation. In his seminal text *Simulacra and Simulation*, Baudrillard provocatively uses the Holocaust as a metaphor to illustrate the effacing of the human with artificial memories, arguing that "artificial memory will be the restaging of extermination" as one passes through "the sound track and image track, through the universal screen and the microprocessor."[1] As a component of artificial memory, the screen enables forgetting; television perpetuates the annihilation of the Holocaust by other means. Baudrillard follows Marshall McLuhan's pronouncement that the "medium is the message" and argues that television is "nothing but a screen."[2] It completely immerses the brain in its own artificiality, its own hyperreality.

The seeming immateriality of the screen represents the process of simulation as it plays out in popular culture. As Baudrillard asserts, "No relief, no perspective, no vanishing point where the gaze might risk losing itself, but a total screen where, in their uninterrupted display, the billboards and the products themselves act as equivalent and successive signs."[3] The total screen signals the death of spectatorial distance and aesthetic illusionism. Distinctions between text and image, man and machine, producers and receivers lose their theoretical import when all can be expressed through the virtuality of the screen. The transformation of all the world into a screen signals the end of aesthetics because the mind is entirely integrated into the hyperreal. The screen is a mirror that reflects reality, which is to say it signifies the incursion of hyperreality in postmodern society. The screen resurfaces in other writings of Baudrillard, as well. In *America*, Baudrillard describes American reality, "It was there before the screen was invented, but everything about the way it is today suggests it was invented with the screen in mind, that it is the refraction of a giant screen . . . as if everything were carried along by, and haloed in, the gleam of the screen.[4] The screen determines events in the everyday, creating a "mental configuration and overall perception" different from a European sensibility where things remain substantial rather than transient.[5]

The essay "Xerox and Infinity" describes the subjectivity of the "Telecomputer Man" amidst the onslaught of intelligent machines.[6] Telecomputer Man interfaces with the machine through the screen that "transforms the process of relating into a process of commutation between One and the Same . . . otherness is surreptitiously conjured away by the machine."[7] The screen offers transparency and "pure" communication, while (presumably real) human interaction functions through the production of difference and otherness that enables contradiction, contemplation, and communication. Baudrillard continues:

> The screen of our images, the interactive screen, the telecomputing screen, are at once too close and too far away: too close to be true (to have the dramatic intensity of a stage)—and too far away to be false (to embody the collusive distance of artifice). They thus create a dimension that is no longer quite human, an eccentric dimension corresponding to the depolarization of space and the indistinctness of bodily forms of expression.[8]

Put another way, the screen seamlessly transforms the real into the sameness of hyperreality where no freedom or choice exists.[9] In *Impossible Exchange*, Baudrillard describes the reality of pure speculation and simulation as the "world of the total screen."[10] The total screen enables a world without otherness where systems converge into a hyperreality in order to escape uncertainty; "total equivalence, total screen, final solution."[11] Again, Baudrillard alludes to the systematic genocide during the Holocaust in conjunction with his mention of the total screen to illustrate how simulation exterminates difference.

Baudrillard's theory of simulation attempts to explain how information and communication technologies displace what has historically constituted reality.[12] The concept of the total screen represents what can be called the "fascist" cultural politics of that simulation. Although a substantial portion of this theory is developed in the last decades of the twentieth century before the mass popularization of computing technologies, it continues to play a pivotal role in new media studies. But for the emerging field of game studies, simulation implies something quite different. More modest in scope and pragmatic in application, a simulation is a computational model of a system and/or environment that enables real-time user interaction by anticipating human behavior.[13] A computer simulation models complex systems as *processes* rather than physical structures.[14] This definition echoes Baudrillard in that simulations are productive, meaning that they produce a particular reality (or hyperreality). As a simulation, a video game presents a simulated reality and provides the means through which a player can navigate through it. In several recent video games made for next generation gaming platforms like the PlayStation 3 and XBox 360, the screen plays a key role in the navigational process. Similar to *Dead Space*, the video games *Mass Effect 2, Splinter Cell: Conviction*, and *Heavy Rain* integrate screens directly into the gameplay. The player in these games, all single-player narratives, use screen technologies to navigate through the simulated realities of the virtual world. How does the presence of the screen in these games compare to Baudrillard's concept of the total screen used to represent the hyperreal?

Returning to *Dead Space*, the layers of screens in the opening sequence resonate with other types of interactive displays. A simulated map provides schematic information about the ship under attack, informing the player about what sections of the ship are operable and open to explore. In elevators and passageways, "entrance" and "exit" signs are animated and onscreen. The glowing text draws attention, making it easier for the player to locate where Isaac needs to go. Transparent screens on the door display information about whether or not it is locked, and above the doors are horizontally oriented screens informing the player where it leads. These screens draw the player throughout the ship to face the threat offered by the Necromorphs, a threat that intensifies throughout the course of the game. The player encounters another screen interface with the "agi-link" or save point. As Isaac approaches, metal tablets unfold and a screen contains links for saving progress. Less an "uninterrupted display . . . [of] equivalent and successive signs" that composes the total screen, the use of screens in *Dead Space* transform the background into maneuverable space within the narrative architecture of the game.[15] Henry Jenkins explains the concept of "narrative architecture" in game studies, "Game designers don't simply tell stories; they design worlds and sculpt spaces."[16] The game creates an "immersive narrative experience."[17] Narrative architecture refers to the way these types of immersive narrative experiences are built into game spaces, and the background screens in *Dead Space* serve as design elements in its construction.

In the hybrid shooter and role-playing game *Mass Effect 2*, the player encounters similar types of screens that act as architectural components in the game's narrative. This sequel re-introduces the character of Commander Shepard who, after being killed in an alien attack on his ship, is brought back to life as a part of "Project Lazarus" by an organization called Cerberus to help solve the mystery of disappearing human colonies. In the Citadel, the human city where Cerebus is located, a holographic Artificial Intelligence system called "Avina" provides information about current events in the city. Screens also display the moving "art" images (organic pools and circular designs) along the stairwell and on video boards in a nightclub called "The Dark Star." The moving imagery provides a kind of feeling to the space as the player navigates the world. Additionally, cylindrical screen stations wrapped by what resembles an LED display can be found throughout the Citadel. When Shepard approaches these stations, it triggers personalized advertisements; one features a large face "talking" to Shepard about his need for a loan, knowing he didn't make any money over the last year. The presence of screen technologies in the Citadel cues into the city's "state of the art" urbanity through these interactive advertisements that utilize niche-marketing strategies. They also create a way to fill out the simulation by providing an historical context displayed not through flashback, but as texture in the narrative architecture of the city.

A large portion of the gameplay in *Mass Effect 2* occurs onboard Shepard's ship, the Normandy. As he and his team (all controllable by the player) explore alien colonies, several types of screens on the ship help the player and Shepard navigate through the storyline. As in *Dead Space*, holographic screens mark the doorways of the Normandy, and the ship in *Mass Effect 2* also has a window-screen like the Kellion. However, *Mass Effect 2* takes the interactivity of the onboard screens a step further. During the course of the game, Shepard communicates with the Illusive Man (a government agent) through a holographic three-dimensional screen located in the middle of a conference table onboard. The Illusive Man is always seen sitting in a single chair behind an expansive presumably onscreen view of a bubbling planet. From the perspective of the Illusive Man, Shepard appears as a hologram. This science fictional interface enables the Illusive Man and Shepard to occupy the same place, but not the same space.

In a similar vein, the "galaxy map" allows the Normandy to travel via the screen. On the bridge of the Normandy, Shepard approaches an interactive map on display. As he walks up a few stairs, the viewpoint pans downward and enters the map that is now contained by the frame of the player's screen. The map provides information about important sites in the game and is organized using orbit rings to show distance; the player can zoom in to an orbit system or zoom out to see the entire galaxy. Moreover, he or she controls a model-sized version of the Normandy. Positioning the ship on the map transports the ship to different locations in the virtual world of the game. While traveling through the screen the player still has to be aware of real world problems like the need for fuel. The galaxy map makes space travel, something that often proves tedious in other video games, a tolerable experience. Moving from point to point often takes up time, but the screen as navigational tool reduces the interruption of the gameplay.

Likewise, Isaac's special "screened" suit in *Dead Space* limits the amount of breaks during the actual gameplay. Through holographic screens projected from his suit, the player controls Isaac's ability to move in different directions, open and unlock doors, and punch. When Isaac shoots his weapon, information about the number of bullets in the gun is displayed using the viewfinder that is another holographic screen. If the player is lost, a glowing "line" projected onto the floor from his left hand shows the way. Other information displayed onscreen through the suit includes: a map of the ship, inventory, personal journal, objectives (what needs to be done to complete each section), and library (can play the video and audio collected throughout the game). Isaac uses video capabilities in the suit to link with others on ship (video and audio chat) to provide the player, who can scroll around and look at the semi-transparent video onscreen emanating from the suit in all directions, with information about how to proceed. Similarly, Norman Jayden, one of the playable characters in the mystery game *Heavy Rain*, has a special pair of glasses that enables him to virtually access and manage files about the case of the Origami killer. Norman's projected "office" can be whatever he wants it to be, with floating orbs constructing a visual map to organize information like fingerprint records or past histories of victims, anything that might help solve the puzzle. As with the special suit, the virtuality of the projected office provides the means through which the player navigates the narrative space of the game.

The variety of screens integrated into the narrative architecture of *Mass Effect 2, Dead Space,* and *Heavy Rain* makes the gameplay seem more seamless and continuous. The animated signs and cylindrical screens for personalized advertisements create a sense of depth to the narrative architectures that compose the simulated realities of these video games. The stealth game *Tom Clancy's Splinter Cell: Conviction* utilizes the screen to construct a sense of seamlessness, but by other means. The game presents flashbacks and other cinematics as projections onto Sam Fisher's surrounding environment rather than standalone sequences that interrupt gameplay. Similarly, projected text is flashed onto sites to give the player directions like "Find the Scientist" or "Capture Lesovsky" to help navigate through the game space. Here, the virtual world becomes a screened reality.

The screen interfaces of the galaxy map, Isaac's special suit, and Norman's virtual office alongside the cityscape as projection screen in *Tom Clancy's Splinter Cell: Conviction* allow the onscreen character to interact with his environment in much the same way the actual screen enables the player to interact with the onscreen simulation. The echoing of one screen within the other creates a layering effect, forming a network between player, playable character, and simulated reality. This simulated reality is not total; however, it cannot be separated from the layers of screens that make the simulation possible. The materiality of the screen, in fact, is a reality—opposed to a hyperreality—the player discovers as the game unfolds.

Screens are often central components to the simulated landscapes of video games in the form of interactive billboards and futuristic holograms, or even as more conventional media forms like projection screens, cathode ray tubes, and computer monitors that the protagonist (and subsequently, the player) can interact with while playing. *Dead Space* and *Mass Effect 2*, or more recently *Heavy Rain* and *Splinter Cell: Conviction* use the screen as the primary means through which the main character and the player who controls him interface with their simulated realities. In these examples, the screen offers a way to seamlessly include key pieces of information about the character or plot into the media landscape. This seamlessness makes the entire experience of playing the game more continuous. The player no longer needs to pause in order to make selections, nor does the developer need to interrupt gameplay to provide back story; the material presence of the screen removes the need for these types of interruptions. In these video games, the screen interface that integrates a player into the action of the game likewise appears to immerse him or her within the hyperreality of the simulation. Yet, this possibility of total immersion—a complete absorption of the body into the simulated environment or the fantasy of virtual reality—is challenged by the materiality of the screen itself as it appears in contemporary American popular culture, a materiality that Baudrillard's concept of the total screen does not anticipate.

Although not specifically referring to the screen, Jay Bolter and Richard Grusin offer a theory of new media that takes this materiality into account. Similarly influenced by McLuhan's media ecological approach in *Remediation: Understanding New Media*, they outline the process of remediation by which older media are refashioned and reused by newer media. The "double logic" of remediation addresses the contradiction between hypermediacy and immediacy in popular culture; "Our culture wants both to multiply its media and to erase all traces of mediation: ideally, it wants to erase its media in the very act of multiplying them."[18] While immediacy seeks complete immersion and erasure of the medium, hypermediacy recognizes the heterogeneity of mediated space by making the viewer aware of the medium as a medium.

The double logic of remediation exposes Baudrillard's assumption that an already established reality exists apart from the processes of mediation, just like he assumes some essential humanity exists apart from the machine. Within the Baudrillardian framework, mediation (especially mass media positioned in stark contrast to the artistry of the theater or literature) destroys reality. However, for Bolter and Grusin, mediation contributes to the construction of the real. The recognition of the materiality of the screen asserts that one's experience of the real is always already mediated, and the cultural affair with hypermediacy exposes this fact. Interactions with the simulation must always occur through some type of material screen, be it the interactive galaxy map of *Mass Effect 2* or the player's television that frames the static snow during the opening sequence of *Dead Space*. Iterating the logic of hypermediacy, the various screens in these video games assert their own reality even as they create a sense of seamlessness and spatial continuity between player and medium. In contrast, Baudrillard's theory of simulation depends entirely on the transparent immediacy of media represented by the immateriality of the total screen.

While more and more screens enter into the narrative architectures of video games, the screen is never total in the way Baudrillard describes it because the interface never disappears entirely. The screen signals difference, illustrated here by my descriptions of several types of screen interfaces, as much as it might signify the sameness of hyperreality. Likewise, mass culture of the twentieth century has given way to a multiplicity of niche markets that commodifies otherness rather than erasing it. The proliferation of screens is part of this process, making digital content including video games available through a variety of screen interfaces be they handheld devices or larger-than-life video boards. Baudrillard's concept of the total screen recognizes a deep-seeded fear of change in the name of conformity that mass and new media often represent. While the materiality of the screen cannot assuage this fear, it indicates the present formation of a continuity rooted in the potentiality of the multiplicity currently being negotiated in popular culture within the simulated and screened realities of video games.

Notes

1. Jean Baudrillard, *Simulacra and Simulation*, trans. Sheila Faria Glaser (Ann Arbor: University of Michigan Press, 1994), 51-52.
2. Ibid., 53.
3. Ibid., 75.
4. Jean Baudrillard, *America*, trans. Chris Turner (London: Verso, 1988), 53-54.
5. Baudrillard, *America*, 54.
6. Jean Baudrillard, "Xerox and Infinity," in *The Transparency of Evil: Essays on Extreme Phenomena*, trans. James Benedict (London: Verso, 1993), par. 6.
7. Ibid.
8. Ibid., par. 9.
9. Ibid., par. 17.
10. Jean Baudrillard, *Impossible Exchange*, trans. Chris Turner (London: Verso, 2001), 5.
11. Ibid., 15.
12. Sean Cubitt, *Simulation and Social Theory* (London: Sage, 2001), 1.
13. Marc Prensky, *Digital Games Based Learning* (New York: McGraw-Hill, 2001), 211.
14. Martin Lister et al., *New Media: A Critical Introduction*, 2nd edition (New York: Routledge, 2009), 42.
15. Baudrillard, *Simulacra and Simulation*, 75.
16. Henry Jenkins, "Game Design as Narrative Architecture," accessed July 4, 2010, http://web.mit.edu/cms/People/henry3/games&narrative.html, par. 3.
17. Ibid., par. 5-6.
18. Jay David Bolter and Richard Grusin, *Remediation: Understanding New Media* (Cambridge: MIT Press, 2000), 5.

Chapter 12

Seriously Fun: Marketing and the Gaming Experience of Nolan's *The Dark Knight*

Lian Amaris

Part alternate reality game and part marketing campaign, the ads and events surrounding the premiere of Christopher Nolan's *The Dark Knight* used groundbreaking experience design to create a virtual Gotham City for real-world interactivity. While alternate reality games have tended to be more of a subcultural phenomenon, taking considerable work to find the point of entry known as "the rabbit hole," Nolan's *The Dark Knight* campaign infiltrated the mass consciousness easily: a simple Google search of the film led to a site called ibelieveinharveydent.com wherein Gotham City's online presence was established. Over time, it was revealed that over forty individual websites for Gotham City companies, stores, and political campaigns were officially used in the game, with several fan sites also contributing to Gotham City's online presence. Once the game began, several sites affiliated with the Batman character of the Joker (including the site whysoserious.com) revealed puzzles, riddles, and instructions to be performed in the real world; additionally, game participants received emails, phone calls and text messages from "Gotham City residents." As people played the game, they were rewarded with teasers, trailers, movie tickets and event passes for *The Dark Knight*. *The Dark Knight* campaign/game acted as an exemplar of a successful commodification of "experience" blurring the lines between consumer and game-player. Additionally, *The Dark Knight* campaign proved that successful marketing can and should utilize a hybrid of the simulated with the "real." Through the use of ubiquitous small screens of cell phones, laptops and PDAs as portals to Gotham, *The Dark Knight* campaign attracted moviegoers to the large screen in staggering numbers.

42 Entertainment (42E), responsible for such notable alternate reality games as *The Beast* (a tie-in with the movie *A.I.: Artificial Intelligence* after it was released) and *I Love Bees* (a campaign prior to the release of *Halo 2*), won the 2009 South By Southwest award for game design for *The Dark Knight* campaign called *Why So Serious?*. In their statements about the work they do, they often refer to "eventizing" the web—their phrase for their underlying mission of amplifying web-based elements in the real world. For *Why So Serious?*, 42E utilized just about every possible environment: Web 1.0 (websites, email); Web 2.0 (Google Maps, Wikis, Blogs, Videos, Social Networking); SMS/text messaging; phone numbers and voicemail messaging systems; large-scale performance events; and small-scale artifacts.

The premise behind the *Why So Serious?* game was to fill in the fictional timeline in Gotham City between the end of the movie *Batman Begins* and the beginning of the upcoming release *The Dark Knight*, and to generate interest in the latter movie through an immersive experience. Within the fictional world of Gotham City, citizens set up a website (ibelieveinharveydent.com) to support Dent's election to the position of Gotham City District Attorney; as a response to this website, ibelieveinharveydenttoo.com was launched as a parody site, which eventually revealed the Joker's involvement in and authoring of a series of games, puzzles, and tasks he would ask people to play to show their allegiance to him. And thus began an epic battle (and the alternate reality game) between the Joker and the rest of Gotham City—whether the Joker was antagonizing citizens, city officials, Batman or other crime bosses. Real-world players of the game participated in a number of real-world events, were recruited via email, phone call, and text message to the sides of Dent or the Joker, were given puzzles to solve, and were rewarded with promotional materials, photos and posters, movie trailers, newspapers, cellular phones and movie tickets.

Marketing Strategies: On Designing an ARG

In *Requiem for the Media*, Baudrillard "analyze[s] and condemn[s] the media as the institution of an irreversible model of communication without a response."[1] Baudrillard's critique of media is that they produce a one-way message to which the masses are subjected without the opportunity to engage in the conversation. However, I argue that 42E's efforts to create their *Dark Knight* marketing experience (and more generally, alternate reality games used in marketing) can be analyzed as a corrective to Baudrillard's critique. I propose that the gaming experience of *The Dark Knight*, though ostensibly a media/marketing venture, enabled players to converse through the media rather than simply receiving the media. Ed Sanchez, a member of the production team responsible for the marketing surrounding *The Blair Witch Project,* said in an interview,

> What we learned from *Blair Witch* is that if you give people enough stuff to explore, they will explore. Not everyone but some of them will. The people who do explore and take advantage of the whole world will forever be your fans, will give an energy you can't buy through advertising . . . It's this web of information that is laid out in a way that keeps people interested and keeps people working for it. If people have to work for something, they devote more time to it. And they give it more emotional value. [2]

In an attempt to elucidate the design of such a media experience, Jordan Weisman, creative director and co-founder of 42Entertainment, explained basic principles of alternate reality game (ARG) design; included in these tenets were the concept of the hive mind, web-based narrative, hiding things in plain sight, and creating what he called an "electronic sphere."

The hive mind of game players arises from the networked nature of the target demographic (18 to 35-year-olds) once they are given a common storyline in which they can emotionally invest, according to Weisman. He argues the common history gives players an impetus to seek each other out. For 42E's first ARG, *The Beast,* Weisman assumed there would be a few hundred thousand players with specific skill sets. However, when millions of discreet players began playing the game, the game designers realized that as a group, the players had unlimited resources, skill sets, and knowledge bases.[3] Because of this, there were no limitations on designing experiences—players as a whole were willing to do anything and everything because they weren't doing everything by themselves, rather, they were working together.[4] Guy Debord, in his work *The Society of the Spectacle,* argues, "Spectators are linked only by a one-way relationship to the very center that maintains their isolation from one another."[5] However, in the case of ARGs developed as campaigns, would-be spectators are transformed into active participants, or game players, and that transformation does not allow for isolation or passivity. On the contrary, it necessitates communication and collaboration and acts as a corrective to Debord's critique.

Weisman explains that part of smart ARG design is developing a narrative structure that is organic to the web—in other words, narrative structure that goes beyond just mutually exclusive audio, image, video or written word. The project of the ARG is to embrace the "chaotic and frustrating nature of the web," wading through all of the unnecessary or irrelevant information to find something meaningful.[6] Weisman is not critical of this aspect of gathering information from the web, but rather, sought to use it strategically for designing ARGs. Weisman states, "I likened it to archeologists who go through a lot of sand looking for a piece of pottery. After they find that shard they have an idea of how to find more, and if there are enough shards they can reconstruct the society that made the pottery."[7] In this way, the game designers work backwards from a completed character-driven emotional story and create all of the evidence that would have existed if the story had been true. The designers then remove the story and hide the evidence inside of thematized puzzles that would exist within the world they have created. The hive mind finds the pieces, gathers the evidence, and begins reassembling the story. As the world is built by the players over time, they are able to distill their wide range of theories into a cohesive idea of the world and what's going on in it. Weisman suggests "because it's moved through the filter of these millions of minds—it's now a personal piece."[8] This sentiment echoes Sanchez's sentiments regarding the success of *Blair Witch* marketing.

The aspect of hiding elements of the game in plain sight is also a strategy outlined by Weisman. According to Weisman, because the target demographic of 18 to 35-year-olds has grown up in an oversaturated mediatized environment, they have developed very sophisticated methods of avoiding marketing; he states, "as a rule of thumb, the bigger the neon sign the faster they'll run the other way. So the premise here was, instead of shouting, go the opposite way and whisper—hide it."[9] As game-players find the messages, they have ownership over the discovery experience and want to talk about it with other people—thus begins the viral campaign. Rather than being bombarded with and then avoiding information about a product (in this case, a movie), consumers (in this case, game players) are actively seeking out more information—a marketer's dream. Weisman says, "It shifts entertainment presentation from exhibitionist to voyeuristic . . . and turn[s] those other media elements from 'must be avoided' into 'must be dissected.'"[10]

The final tenet outlined by Weisman is to surround the target audience in an "electronic sphere."[11] Knowing that the people are constantly traveling with their communications networks virtually at all times, any ARG campaign should take full advantage of that traveling network. Therefore, no campaign ever lives solely online, but, as I outline in the case of *Why So Serious?*, uses every kind of media access possible. For Weisman, the ideal scenario is one where the consumer/player can immerse themselves in the product/game as much as twenty-four hours a day—with a symbiotic reaching out to the game and having the game reach out to them.[12] It is important to note that Weisman's electronic sphere is not actually created by the ARG, but rather is an environment that already exists, of which the ARG takes advantage. To further explore this environment, I look to Jenkins' concept of "convergence culture" and de Souza e Silva's concept of "hybrid space."

Ubiquitous Platforms for Gameplay and Game Space

In *Convergence Culture: Where Old and New Media Collide*, Henry Jenkins addresses changing responses to media which he calls "convergence." Choosing not to focus on the devices or the technology itself, but rather the cultural response to it, he posits, "convergence represents a cultural shift as consumers are encouraged to seek out new information and make connections among dispersed media content. This book is about the work—and play—spectators perform in the new media system."[13] Though Jenkins is not directly addressing ARGs, his inclusion of "play" articulates the connection between ubiquitous devices and the potential for play. Additionally, though not central to his argument, his claim that spectators are *performing* within this space reveals the relationship between play, performance, action, and liveness, which I will address in the following section. Furthermore, Jenkins' concept of convergence culture is an active response to media and addresses aforementioned postmodernist critiques of said media:

> The term *participatory culture* contrasts with older notions of passive media spectatorship. Rather than talking about media producers and consumers as occupying separate roles, we might now see them as participants who interact with each other according to a new set of rules that none of us fully understands. [14]

The participatory nature of ARGs speaks directly to this active (rather than passive) mode of engagement, and is made possible through ubiquitous devices. It may provoke some anxiety to think of a marketing campaign as authoring "participatory culture," and perhaps rightly so, as Jenkins' addresses the previously-named "spectator" now as "consumer." Though the transformation of spectator to consumer is beyond the scope of this article, my analysis of ARGs can never be entirely separate from marketing goals. In his account of the progression of video game culture, Nate Garrelts traces the development of arcade games to in-home game consoles to games on ubiquitous devices as a response to consumer-driven demand. He accurately argues, "as new hardware is introduced into the market, it is quickly repurposed to play games—the most recent examples are PDAs and cell phones, both of which are increasingly popular gaming platforms."[15,16] He goes on to state, "Perhaps more importantly, the existence of multiple gaming platforms has been the catalyst for economic success and innovation."[17] In other words, Jenkins' convergence of media is a result of consumer desire for more and multiple access points to content, and in this case, specifically game content; ARGs respond innovatively to that desire for more content in an active, participatory way, utilizing ubiquitous devices to support and enhance game play. Jenkins' articulation of media convergence references anthropologist Mimi Ito's research on Japan's culture of "media mix":

> On the one hand, the media mix strategy disperses content across broadcast media, portable technologies such as Gameboys or cell phones, collectibles, and location-based entertainment centers from amusement parks to game arcades. On the other, these franchises depend on hypersociability, that is, they

> encourage various forms of participation and social interactions between consumers.[18]

Again, this notion of hypersociability—often seen as counterintuitive considering the potential isolation of personal computers and devices—is overtly present in ARGs, especially those that require group problem-solving skills, or employ large-scale public events and performances. Furthermore, Ito's inclusion of amusement parks and game arcades as sites of potential content reinforces the twin elements of fun and play, inherent within successful ARGs, and furthers the connection to public spaces that are often utilized. Adriana de Souza e Silva explores the effect ubiquitous devices have on public spaces, and renames that space "hybrid space."[19] I propose that Weisman's successful "electronic sphere" necessarily makes use of this hybrid space.

De Souza e Silva suggests, "[h]ybrid spaces merge the physical and the digital in a social environment"[20] and are "created by the constant movement of users who carry portable devices continuously connected to the Internet and to other users."[21] De Souza e Silva is careful to make the distinction between "virtual" spaces and the hybrid space created by portable devices; for the purposes of her argument, users are not necessarily concerned with immersion in a virtual world, but rather with connecting to friends or physical space through location awareness.[22] This concept of the hybrid space is complicated and enhanced by the ARG player who utilizes portable devices with the specific intention of immersion into an "alternate reality." ARGs necessarily straddle the virtual-real divide in that they exist on any available platform—including those online or in the physical world. Moreover, as the dimensions of public social spaces are reconfigured by ubiquitous portable devices, participation in an ARG adds yet another layer to a Foucauldian heterotopia.

Just as Garrelts traces the shift from game arcades to home consoles to portable devices, de Souza e Silva articulates a similar trajectory with regards to space: "Once there is a shift from static to mobile interfaces, users are no longer required to sit in front of their computers, but rather, they move around in urban spaces—which are already social public spaces."[23] But here I must offer the important rhetorical note that users are not simply going to public spaces, but rather portable devices allow for a *return* to public space, after the isolation created by personal computers.[24] I would like to suggest that through the use of portable devices, ARGs facilitate the return to public space wherein live, physical interactions appeal to ARG participants and offer the elusive antidote to isolation and passivity. Because ARG players are connected in real-time and are participating in events that occur in the physical world, their traversal of hybrid space calls attention to the tenuous relationship between liveness and mediatization, as articulated by Philip Auslander.

Between Liveness and Mediatization

According to Jenkins, convergence culture "is about the work—and play—spectators perform in the new media system."[25] Therefore I would like to turn our attention to performance in a mediatized culture. Historically, marketing campaigns relied on broadcasting information, rather than on one-time performance events, if only because broadcast media are more cost-effective and reach a greater audience. Paradoxically, a significant element of *Why So Serious?* was the incorporation of events and performances that occurred live in public spaces including: campaign rallies, skywriting, and events in bakeries and bowling alleys. Additionally, mediatized performance events included breaking-news stories shown for Comcast cable customers, teasers in the form of video and audio clips, and phone messages enhancing game play. Similar to their utilization of hybrid spaces, ARGs intentionally make use of the relationship between live and mediatized performance. In *Liveness: Performance in a Mediatized Culture*, Philip Auslander states, "I am treating live and mediatized performance as parallel forms that participate in the same cultural economy"[26] and I would argue that this cultural economy is indeed the site of success for any ARG.

> From ball games that incorporate instant replay screens, to rock concerts that recreate the images of music video, to live stage versions of television shows and movies, to dance and performance art's incorporation of video, evidence of the incursion of mediatization into the live event is available across the entire spectrum of performance genres. This situation has created an understandable anxiety for those who value live performance, and this anxiety may be at the root of their need to say that live performance has a worth that both transcends and resists market value.[27]

The ARG (as hybrid genre that incorporates theater, play, narrative, design, and marketing) provokes but perhaps assuages some of this anxiety precisely because ARGs make use of every available platform—including the live event; this is made possible by ubiquitous portable devices. A symbiosis occurs wherein the live event is promoted or revealed through online puzzles, and the game world is then enhanced by the live event. Auslander's acknowledgment of the market value of live performance echoes the sentiments of Jenkins, Garrelts and even Weisman, and identifies "the anxiety that underlies many performance theorists' desire to reassert the integrity of the live and the corrupt, co-opted nature of the mediatized."[28] Though it may seem self-evident that a marketing campaign cannot possibly resist cooptation, corruption, or the market, I wish to suggest that if there is a "depreciation of live presence,"[29] perhaps ARGs can ameliorate that depreciation precisely because they utilize live performance as another medium to "surround [players] and allow them to immerse themselves."[30] As with the relationship between performance and media, the long relationship between performance and play, reaching back to prehistoric humans, is not without its own anxieties, some of which I will address.

Serious Players and Dark Play

The very first puzzle of 42E's game *Why So Serious?* was the first reveal of the Joker's face in May 2007. It began with the website ibelieveinharveydent.com which featured a campaign poster of "Dent for District Attorney." Shortly after the site was launched, a second site was found, partly with the help of joker playing cards that were found in a comic book store in Southern California. ibelieveinharveydenttoo.com featured the same campaign poster, but was "jokerized" with black eyes, and a red smile, in what would become a calling card of the Joker for the entire campaign. There was a place to enter an email address, and those who did received an email from the Joker himself, instructing them to enter coordinates into a previously hidden part of the site. Those coordinates corresponded to pixels, and as more and more pixels were entered, a picture was revealed behind the campaign poster—it was the first image, a photograph, of the Joker looking straight out to his new audience, frighteningly disfigured and covered in grotesque makeup. His face acted as a provocation, posing the very question he asks throughout the movie just before attacking his victims— "do you want to know how I got these scars?"

Marshall McLuhan aptly points out, "There is always a tingling sense of oddity and fun in play or games that renders the very earnest and very serious person or society laughable."[31] This dichotomy of play versus seriousness is the very paradigm within the character of the Joker functions. The name of his website "Why So Serious" functions by first asking the actual question, and then providing a corrective to what the Joker sees as the problem with society (i.e., seriousness) through games, chaos, and reckless fun. McLuhan's "very earnest or very serious person" here is clearly Harvey Dent, the Joker's rival. Caillois, in *Man, Play and Games*, reminds us that true play cannot be forced onto its players, that all play must be voluntary.[32,33] In the actual film, the Joker's enjoyment derives from forcing people to play his games; however, in the ARG, all players are presumably voluntary, and are able to participate in whatever ways they desire. So the ARG functions as a set of games-within-a-game, wherein real-world participants play the ARG designed by 42E, but inside the ARG are the games the Joker has designed for Gotham citizens to play. Jesper Juul reminds us that the most important element of a game (as different from play) is that there are rules—the rules form the skeleton of the game and define the game space as different from the regular world space;[34] Huizinga calls this a "magic circle."[35] Paradoxically, again the Joker undermines our understanding of games, when he forgoes rule sets, and clings to chaos. Indeed one of his games results in the ransom note-style statement (and line from the film), "the only sensible way to live in this world is without rules." Though chaos is the antithesis of a game as it relies on subverting, undermining, and eliminating rules altogether, these elements are clearly present in Richard Schechner's articulation of "dark play":

> Dark play involves fantasy, risk, luck, daring, invention, and deception . . . Dark play subverts order, dissolves frames, and breaks its own rules—so much so that the playing itself is in danger of being destroyed, as in spying, double agentry, con games, and stings. Unlike carnivals or ritual clowns whose inversion of established order are sanctioned by the authorities, dark play is truly subversive, its agendas always hidden. Dark play rewards its players by means of deceit, disruption, and excess.[36,37]

However, even as the Joker fluctuates between chaos and rules sets, the puppet masters of 42E keep the structure of the game intact. In many ways, the Joker serves as an embodiment of what Douglas Rushkoff says of cyber culture in general: he is a "reminder that the rules are not fixed and that those who recognize this fact have the best time."[38]

Conclusion

The first puzzle which revealed the Joker's face marked the entrance to a game that included everything from sky-written teasers, to cellular phones baked into cakes, to a real-world voter's campaign led by the "Dentmobile," to episodes of "Gotham Tonight" which aired on television for Comcast customers. Game players were asked to take photographs of themselves dressed up as the Joker and submit them to his website and alternately asked to assemble at rallies in support of Harvey Dent. All of these events amplified immersion into the world of Gotham City, and fortunately for 42E, product awareness arose through game play, making the marketing campaign hugely successful and the movie itself much anticipated. Baudrillard claims in "The Implosion of Meaning in the Media" from *Simulacra and Simulation*,

> We live in a world where there is more and more information, and less and less meaning . . . Information devours its own content. It devours communication and the social . . . Rather than creating communication, it exhausts itself in the act of staging communication. Rather than producing meaning, it exhausts itself in the staging of meaning.[39]

Though communication and meaning are indeed both staged through a marketing campaign, perhaps ARGs do not entirely devour—with a red, artificial smile—communication and meaning. Alternately, meaning is produced through experience, bolstered by active communication, and the social is made paramount. Rather than promoting the passive consumption of information, ARGs offer meaningful, experiential connections to popular cultural events such as movies. And while there is always an inescapable bottom line, perhaps such marketing campaigns give more by way of experience than they take when they ask you to buy their product. There is always a product that must be sold, but perhaps the exchange rate on fun is worth the price. And so to aggressive marketing firms and anti-advertising media critics alike, I ask, why so serious?

Notes

1. Jean Baudrillard, *Simulacra and Simulation* (Ann Arbor, MI: The University of Michigan Press, 1994), 84.
2. Henry Jenkins, *Convergence Culture: Where Old and New Media Collide* (New York, NY: New York University Press, 2006), 105.
3. Jordan Weisman, "42's 5 principles of arg." *CREATIVITY*, Accessed July 2, 2010, http://creativity-online.com/news/42s-5-principles-of-arg/106069.
4. "42's 5 principles of arg."
5. Guy Debord, *The Society of the Spectacle* (New York, NY: Zone Books, 1994), 22.
6. "42's 5 principles of arg."
7. "42's 5 principles of arg."
8. "42's 5 principles of arg."
9. "42's 5 principles of arg."
10. "42's 5 principles of arg."
11. "42's 5 principles of arg."
12. "42's 5 principles of arg."
13. Jenkins, *Convergence Culture*, 3.
14. Jenkins, *Convergence Culture*, 3.
15. Nate Garrelts, *Digital Gameplay: Essays on the Nexus of Game and Gamer*, ed. Nate Garrelts (Jefferson, NC: McFarland & Company, Inc, 2005), 6.
16. I would add the iPad to this list.
17. Garrelts, *Digital Gameplay*, 6.
18. Jenkins, *Convergence Culture*, 112.
19. Adriana De Souza e Silva, "From Cyber to Hybrid: Mobile Technologies as Interfaces of Hybrid Spaces" in *The Cybercultures Reader*. 2nd Ed, edited by David Bell and Barbara M. Kennedy (New York, NY: Routledge, 2007), 757.
20. De Souza e Silva, "From Cyber to Hybrid," 759.
21. De Souza e Silva, "From Cyber to Hybrid," 758.
22. De Souza e Silva, "From Cyber to Hybrid," 759.
23. De Souza e Silva, "From Cyber to Hybrid," 765.
24. It is also important to note that the public gathering place of Ancient Greece, the *agora*, began as a public assembly but became a public marketplace. This transition is echoed in the contemporary spectator-turned-consumer.
25. Jenkins, *Convergence Culture*, 3.
26. Philip Auslander, *Liveness: Performance in a Mediatized Culture* (New York, NY: Routledge, 1999), 5.
27. Auslander, *Liveness*, 7.

28. Auslander, *Liveness*, 39.

29. Auslander, *Liveness*, 36.

30. "42's 5 principles of arg."

31. Marshall McLuhan, *Understanding Media: The Extensions of Man* (Cambridge, MA: The MIT Press, 1994), 244.

32. Roger Caillois, *Man, Play, and Games* (New York, NY: Schocken Books, 1961), 10.

33. Juul takes direct issue with this imperative when he argues, "Is it not a game if social pressure forces the player to play? Because human motivation is too complex to be simply explained in terms of its being voluntary/involuntary, I believe that it is not possible to meaningfully describe whether games are voluntary or not" (31-33). Perhaps with regards to the Joker's games, it is not whether or not they are voluntary, but whether or not the potential outcome of death is "fun" for the player.

34. Jesper Juul, *Half-Real: Video Games Between Real Rules and Fictional Worlds* (Cambridge, MA: The MIT Press, 2005), 36-38.

35. John Huizinga, *Homo Ludens: A Study of the Play Element in Culture* (Boston, MA: Beacon Press, 1950), 20.

36. Schechner, Richard, *Performance Studies: An Introduction.* 2nd Ed. (New York, NY: Routledge, 2006), 119.

37. Pertinent with regards to the persona of the Joker, Schechner also identifies costuming or disguise as inherent in dark play. "Assuming a new or alternative identity, even briefly, is very important. Masking, cloaking one's ordinary self just to get away from the humdrum, is also important . . . In disguise play . . . alternative selves are given license" (121).

38. Rushkoff, Douglas, *Cyberia: Life in the trenches of hyperspace* (Manchester: Clinamen Press, 2002), 158.

39. Baudrillard, *Simulacra and Simulation*, 79-80.

Chapter 13

Transgressive Simulation: Violence and Reality in Extreme Championship Wrestling

Benjamin D. Hagen

In this essay, I hope to earn the concept "transgressive simulation," a neologism that targets the relationship between reality and violence, two notions of great import to the study of art, literature, film, and other performative media. In the first half, a brief encounter with Jean Baudrillard opens up the possibility for this concept, raising the question of how one can call a simulation "transgressive" in the midst of an era where one can neither rediscover "an absolute level of the real" nor recognize an illusion with any degree of certainty.[1] In order to attend to this problem, I target professional wrestling as an exemplary worksite for thinking through what I believe to be a chiastic relation between violence and reality. In particular, I focus on Extreme Championship Wrestling (ECW), an "underground" promotion that operated throughout the 1990s, and argue that the spectacle of professional wrestling as manifested in ECW serves as a fascinating example of transgressive simulation insofar as its promoters, bookers, and wrestlers fold self-consciously violent methods and strategies into their shows. Such methods produce dramatic, yet imperfect, ruptures in the precession of simulacra, which arrest spectators and make possible an affective encounter with the mobility of and the interplay between reality and violence.

Two Bank Heists: Hypersimulation and Transgressive Simulation

Near the middle of his essay "The Precession of Simulacra," Baudrillard follows his analyses of religion, ethnology, Disneyland, and Watergate with a challenge that targets the problem of recognizing the illusoriness of illusions. He dares us to "[o]rganize a fake holdup . . . [that] remain[s] close to the 'truth,' in order to test the reaction of the apparatus to a perfect simulacrum."[2] Baudrillard assures us that the result of such a prank would be devastating: "a policeman [would] really fire on sight; a client of the bank [would] faint and die of a heart attack; one [would] actually pay . . . the phony ransom."[3] The adverbs "really" and "actually" carry great theoretical weight in this passage because they perform and portray Baudrillard's point that violence, even if only threatened and/or perceived, serves as a *medium* or channel for an eruption of a felt reality that overwhelms the phoniness of the ransom. The intrusion of a sensible actuality—here a sensation of approaching death—makes the "fake holdup" resonate violently with effectual reality, rendering "submission and transgression equivalent."[4] According to Baudrillard, the blurring of this binary opposition is dangerous because the law bases itself upon our ability to rationally differentiate between submission and transgression. Since the hypothetical policeman cannot account for the trickery of the "hypersimulation,"[5] he will answer the perceived violence as if it were transgression; "opt[ing] for the real"—a fascinating phrase—he will shoot.[6]

Prank shows such as *Candid Camera* or MTV's *Punk'd* serve as "real life" examples of Baudrillard's hypersimulation.[7] Both shows "remain close to the truth," as he insists, capturing real reactions to imagined situations. In addition, stations broadcast these pranks and arranged catastrophes, effectively popularizing the horror of one or a few and transforming it into the entertainment of many. How might one account for the fascination that these programs (and others like them) inspire? Perhaps we have, as Roland Barthes writes in "The Discourse of History,"

> a taste for the reality effect, attested to by the development of specific genres such as the realistic novel, the private diary, documentary literature, the news item, the historical museum, the exhibition of ancient objects, and, above all, the massive development of photography, whose sole pertinent feature . . . is precisely to signify that the event represented has *really* taken place.[8]

A structurally integral, but variable, *safe distance* grounds the "taste" that Barthes describes here. Just as readers of "the realistic novel" often remind themselves that the characters and their pains or pleasures are not real while simultaneously experiencing or imagining these characters *as if* they were, television audiences may rest assured that the fictive cruelties of *Candid Camera*'s pranks will eventually resolve themselves without heart attacks or bullets by virtue of the very fact that such pranks *are televised.* ("No one would air a prank gone wrong," we might tell ourselves.) Once the host reveals that the arranged catastrophe is nothing but a scripted joke, panic settles into manageable responses: laughter, anger, annoyance, or relief.

The reaction of the participants before this revelation, however, matches the responses of the imagined victims of Baudrillard's bank-heist. Those hypothetical characters and these participants perceive something real *without* the safety of the reality-effective medium that audiences enjoy; they are not privy to the illusoriness of this particular illusion. Though the pretense of *Candid Camera* is really harmless, actual harm inhabits its premise because its victims react *as if* harm were really being done. Unlike Barthes's list of reality-effective media, the violence at the heart of *Candid Camera* and *Punk'd* does not signify "reality"; rather, it brings sensations of a very actual *reality* with it. Like the officer who encounters the perfectly simulated bank robbery, the victims of prank shows more often than not opt for the real. Perhaps this is illustrated when contestants respond, "This can't be happening!" Such self-assurance actually attributes, I argue, something one might call "reality" to sensations of violence. Violence—that assemblage of perceptions and actions—does not serve, then, as a medium for a Barthian reality effect; rather, it ushers in the urgent and often overwhelming rhythms of the real.

I want to offer a second challenge in response to Baudrillard and these hypersimulations: let us stage another holdup, including the officers, the clients, and the tellers in the planning. We might script the action and dialogue, but unlike the original challenge, there is no need to remain "close to the truth." In addition, we will incorporate non-lethal weaponry: guns with rubber bullets; smoke grenades; pepper spray; tasers; etc. Lastly, let us film and broadcast the spectacle.

This revised heist would not be a "hypersimulation" in the *Candid Camera* or *Punk'd* sense. Barring passersby unaware of our script, no one—not spectators nor participants—would "opt for the real." And yet . . . what might one make of the potentially violent disruptions of this game? Of the bruises and contusions from the rubber bullets? What if one hits a client in the eye? Of the genuine fits of coughing, choking, and/or burning caused by pepper spray or a smoke grenade? Of seizures that the tasers might induce? These potential disruptions that inhabit, indeed that are inherent, to this packaged and presented simulation constitute a preliminary sketch of what I call "transgressive simulation": a self-conscious spectacle that announces itself as such *in spite of* and *by way of* complex, excessive, risky, and desperate strategies. Though this heist would initially appear to be a matter of grown ups playing pretend—and thus a *dis*simulation[9]—I argue that such a spectacle would actually and imperfectly interrupt the familiar generation of hyperreality with disturbing and disruptive sensations of a needless violence within the confines of a phony script. Speaking, then, of phoniness and scriptedness, I turn to an exemplary worksite for transgressive simulation: Extreme Championship Wrestling.

This Is Extreme: Challenging Barthes

The public is completely uninterested in knowing whether the contest is rigged or not.
– Roland Barthes[10]

In the ring, and even in the depths of their voluntary ignominy, wrestlers remain gods because they are, for a few moments, . . . the pure gesture which separates Good from Evil, and unveils the form of a Justice which is at last intelligible.
– Roland Barthes[11]

Although fifty years plus have passed since the first publication of Barthes's "The World of Wrestling," the scholarship and criticism on professional wrestling has struggled to rethink the two claims of these epigraphs: (1) that wrestling's scriptedness is irrelevant to its fans and (2) that wrestling maintains and sustains clear ethical or moral divisions. For the most part, critiques that explore wrestling's theatrical side (a theme also developed by Barthes) read the heterogeneous phenomenon symptomatically, focusing primarily on how wrestling supports or resists, reifies or revises Western values.[12] While many critics offer complicated readings of *how* professional wrestling sustains/resists cultural norms, many of these same writers nevertheless assume that the Barthian formulation of a fully intelligible ethics or justice, that is, the unambiguous maintenance between Good and Evil, still applies to this spectacle. Henry Jenkins, for instance, claims that wrestling illustrates "central contradictions within our contemporary constructions of masculinity,"[13] but he accepts that wrestling must be "unambiguous,"[14] a universe of "overstatement" without moral complexity,[15] a "masculine melodrama" seeking "the restoration of moral order, the creation of a just society from a world where the powerful rule."[16] Likewise, in an article concerned with explaining why "heels" (re: bad guys) often win, Campbell argues that these televised serials maintain an entertaining deterrence of an "[u]nequivocal justice" that must and *will* be satisfied, only to be deterred again as other "bad guys" feud with the hero.[17] Earlier in his essay, Campbell confidently writes, "Wrestling fans need not spend any intellectual energy making sense of wrestling."[18]

The trouble, however, with this body of criticism is that it assumes (or chooses to investigate) only one model of professional wrestling—in fact, the *same* model that Barthes studied over fifty years ago in France. As John Rickard points out in his short genealogy of Australian and American wrestling, the spectacle has evolved from bouts with genuine outcomes in the early twentieth century—"even if there had been some understanding to give the crowd its money's worth"—to matches in the 1920s that required more theatrics in order to satisfy the increasing expectations of the audience.[19] This development "from a legitimate sport [to] . . . a morality play and, increasingly, an exotic entertainment" did not, however, climax and/or halt with the rise of nationally telecast federations like the World Wrestling Federation (WWF, now World Wrestling Entertainment [WWE]) and World Championship Wrestling (WCW, which Vince McMahon, chairman of WWE, purchased in 2001).[20] From the early 1990s to 2001, a small organization named Extreme Championship Wrestling emerged, a promotion that bookers, promoters, and wrestlers self-consciously presented as a "revolution" to tamer, more popular brands. I argue that ECW's revision of the popular model of professional wrestling transforms the 1980s and early 1990s spectacle of heroes and villains into a transgressive simulation, thereby disrupting the Barthian foundation of pure intelligibility upon which much, in fact *most*, wrestling criticism grounds itself.

But what exactly *was* extreme about ECW?[21] There are many answers to this question, but a fascinating claim to "authenticity" grounds all of them, that is, an authenticity to which the WWF and WCW had no claim. In other words, while ECW accepted the confines of a "script," the promoters and wrestlers sought to draw in new fans with a "realer" brand of professional wrestling. For instance, one wrestler, whose gimmick was that of a grunge cult-leader, "seduced" both the wife and the seven-year-old son of a second wrestler, troubling the intelligible distinction between "real life" and the world of wrestling. In addition, the shows exposed more skin (primarily female), included more language that would keep censors busy, and flaunted the fixed nature of the spectacle through interviews where wrestlers broke in and out of their television personae (simultaneously introducing their "real me" while exposing the scriptedness of everything else).[22] But ECW's extremity is best demonstrated through a scripted violence that challenges the supposed intelligibility of wrestling, much like my proposed bank heist challenged the intelligibility of the spectacle through the additive of non-lethal weaponry. This violence takes many forms: athletic maneuvers that drop jaws, suplexes and slams that spike performers on his or her head, staged falls from twenty-foot balconies, etc. I'd like to focus, however, (and this is where ECW and my bank heist cross) on ECW's love affair with weapons, that is, on the *perversion of objects* such as broken glass, thumbtacks, spikes, forks, pliers, broken furniture (sometimes furniture that was on fire!), but—most importantly—barbed wire.

Although this last item does play a minor role throughout the history of professional wrestling, this object occupies a space of privilege for this promotion. To illustrate this, one only has to look to the very inscription of the letters *E-C-W*—whether written on signs, interview backgrounds, ring skirts, or VHS/DVD cases. Barbed wire, more often than not, decorates these three letters: either graphically wrapped around or inscribed within them. This is strange iconography, given that this object normally serves as a boundary marker, a visual threat as much as a physical one *against* trespassing, *against* transgression. As a restraint against careless or malicious wandering, barbed wire says to the passerby, "Don't Cross. Not Allowed Here." In ECW, barbed wire remains a warning but not a total restraint. Rather, this imagetext disturbingly invites the wrestling fan. "Be prepared," it says, "to be caught in this wire, to be caught in a threshold. Be prepared to witness something as *real* as these painful barbs." Disturbingly, then, ECW revises the "liminal space" that Dalbir Sehmby attributes to professional wrestling, dramatizing, emphasizing, and overflowing the dangerous "spectator position of marginality" that fans occupy between "the margins of fictional knowing and nonfictional uncertainty" as well as the convergence of "sports and drama, masculine narrative and feminine narrative forms, sexualized and non-sexualized display, and reality and fiction"; this liminal space is strewn or decorated, so to speak, with barbed wire.[23]

"Born to Be Wired": Terry Funk vs. Sabu

But this wire, as I've mentioned, is also a weapon. In ECW's developing years, wrestlers used it to cut themselves and one another during their matches, inciting groans and applause and chants from the crowd—"ECW! ECW! ECW!" The most disturbing use of barbed wire in company history transpired in August of 1997 in a match pitting a wrestler named Sabu (billed as "suicidal, homicidal, genocidal") against 53-year-old Terry Funk for the company's Heavyweight Title.[24] Unlike most wrestling matches, this "contest" saw the three cables or ropes that usually surround the ring taken down and replaced with several dozen yards of barbed wire so tightly wound that both wrestlers could lean all their weight against it without falling to the cement floor below. Although both men begin the match hesitantly, it does not take long for either to get caught in the barbs. Funk and Sabu propel one another into them, press one another's body into them, and whip one another with strands clipped from the longer strands, thus creating scenes of "torture" that Barthes could not have foreseen.[25] As the match progresses, their skin and clothing tears, and Sabu—though the scripted victor—comes away with a punctured vein in his left arm as well as a ripped bicep, an injury that occurs midway through the bout (his manager, Bill Alfonso, fetches a roll of athletic tape with which Sabu tapes the wound, nearly from wrist to shoulder, before continuing).

Much can be said about the disturbing encounter between Funk and Sabu, about the strange and disturbing moments when Sabu neglects to sell the impact of Funk's punches in order to continue taping his arm (thus drawing attention to the phoniness of the punches *as well as* the reality of the open wound), about the hush from the normally rabid crowd when both men become so entangled in the wire that they can no longer separate in order to continue the match. These oscillations between self-conscious scriptedness and the very real cuts, blood, and pained expressions present ECW *as* a transgressive simulation, troubling the confident marketing of its extremity (that is, the promise of a "realer" brand of professional wrestling) and exposing it as a "[p]anic-stricken production of the real"[26] in which eruptions of needless violence—akin to the rubber bullets and smoke grenades of the bank heist—challenge Sharon Mazer's claim that professional wrestling creates "a spectacle of violence that is not actually violent"[27] and encourage worry over Barthes's claim that "wrestling is [not] a sadistic spectacle."[28] Perhaps ECW corroborates Baudrillard's inversion of Borges at the beginning of "The Precession of Simulacra"—an inversion that he dismisses immediately—whereby vestiges of reality, through the special claim of extreme violence on them, do exist, are observed, can be pinpointed (however brief the glimpse) in "*the desert of the real itself*."[29]

But as I have demonstrated, reality and violence are not quite this simple. Like the revised Baudrillardian holdup, ECW's "revolution"—far from returning to any sort of "real" or exiting from the precession of simulacra (it is no more a "real" form of wrestling than the bank heist is a "real" holdup)—actually re-injects anxiety over the fixed nature of wrestling, transgresses the models of professional wrestling that precedes it, dismisses the ethics of Good and Evil in favor of equal-opportunity barbed wire, and thus overturns the theoretical ground upon which nearly all wrestling criticism grounds itself; the barbed wire could care less whether it punctures a hero or a villain. But in a strange way, the gratuitousness of these scenes of torture end up heroicizing *all* performers, both good *and* evil; indeed, the crowds chant in admiration for both Funk and Sabu during and after their match, acknowledging an appreciation for their performative and disturbingly corporeal sacrifices. Above all, then, ECW confronts and sweeps up its fans with the theoretically frustrating mobility of, indeed the chiastic, mutually presupposing relation between, violence and reality. Again and again throughout the history of this promotion, these two sensations prove that the boundaries separating fact from fiction, the fixed from the competitive, truth from the lie, reality from illusion are ever permeable. While the "real life" storylines, schizophrenic interviews, and horrifyingly "real" barbs on the one hand appear to authenticate ECW, the restraints of the game and of the "script" (no matter how loose) create a dissonance, a slowing down of certainty and an intensification of uncertainty that becomes the very substance of what the fans enjoy. Indeed, this substantive violence confronts us even before Sabu and Funk touch one another; it encounters us as and through the supplemental weaponry—the tacks, the fire, the barbed wire—that portray and enact "the irreducible complexity within which one can only shape or shift the play of [presence/violence/reality] or [absence/performance/illusion]."[30] We see something of the logic of the "transgressive" in this notion of Derridean supplementarity, particularly when an object meant to avoid transgression becomes part of the very practice, part of the very calculus, part of the very mechanics of an affective violence that breaks open violence beyond the bounds of representation and that intensifies the rhythms and frequencies of a reality that is brought to a kind of life through this affect.

The novelty of Extreme Championship Wrestling, then, in a way has little to do with the *actual event* of wrestlers cutting one another or of performers shortening their careers by diving from balconies onto concrete floors or through wooden tables. Rather, the novelty of ECW and transgressive simulations in general is that they reframe the encounter with violence, making possible "a practice of response to violence that insists on an encounter with violence on its own terms" and that attempts to avoid the various methods through which we might judge or understand it—e.g., a logic of means and ends or a system of hermeneutics.[31] This insistence on a new mode of response that Marco Abel suggests does not deny the physical toll that the glass and fire and barbs inflict or the stark (un)ethicality of these performances. Rather, to attempt to inhabit the space of the wrestling fan and to encounter these acts of violence without judgment allows one to approach wrestling otherwise, from a space that allows one to see that the relation between audience and performer in ECW, though certainly grounded in the strategic deployment and enjoyment of disturbing excess, does not necessarily exclude a tremendous affect of care, devotion, and loyalty from its transgressive space.

For many fans in an ECW audience, the time frame for the encounter with violence and a related mode of response is very short; for the most part, a fan jarred by the wounds of Funk and Sabu quickly recovers in order to chant, "ECW! ECW!" (For a person *new* to the spectacle, however, this window remains much wider.) But deferring judgment of a transgressive simulation also opens up the possibility for an encounter with what I have called "reality": not an objective image of *truth*, but a rhythm or sensation. In his essay "The Reality Effect," Barthes infers a relationship between the mobile abstractions of violence and reality: "it suffices to recall that, in the ideology of our time, obsessive reference to the concrete"—this can apply to the strategies of excess in the bank heist or in ECW—"is always brandished like a weapon against meaning, as if, by some statutory exclusion, what is alive cannot not signify."[32] In other words, the transgressive simulation, in its modifications of perception and its disturbing desperation to jar rather than conceal, dramatizes this brandishing of the concrete, something in excess of the reality effect, in order to defer meaning, to unveil the detachment of violence and reality from all restraints, to locate them in a liminal space "between," and to train us in reading dissonances that offer critics, audiences, or performers a sense of violence and reality as mobile abstractions that might actually provide food for thought, that might very well teach us something about violence and its role in the entertainment of others, and that might make visible, and thus *real*, new forms of harm that are as of yet invisible to us. This is not to say that one should celebrate the blood-letting and (self-)torture that occurred in ECW—and continues to occur here and now in other companies, particularly in small promotions—but rather that acknowledging and investigating the complexities of the transgressive simulation might very well lead to more rigorous and dutiful reformulations and redefinitions of an ethics of entertainment.

Notes

1. Jean Baudrillard, "The Precession of Simulacra," in *Simulacra and Simulation*, trans. Sheila Faria Glaser (Ann Arbor: University of Michigan Press, 1994), 19.
2. Ibid., 20.
3. Ibid.
4. Ibid., 21.
5. Ibid., 19.
6. Ibid., 21.
7. For those unfamiliar with these programs, free online archives are now available. For *Candid Camera*: http://www.candidcamera.com. For *Punk'd*: http://www.mtv.com/shows/punkd/series.jhtml.
8. Roland Barthes, "The Discourse of History," in *The Rustle of Language*, trans. Richard Howard (Berkeley: University of California Press, 1989), 139.
9. Baudrillard, "The Precession of Simulacra," 9.
10. Roland Barthes, "The World of Wrestling," in *Mythologies*, trans. Annette Lavers (New York: Farrar, Straus and Giroux, 1972), 15.
11. Ibid., 25.
12. I am not accusing this body of criticism of somehow missing the most important dimension of professional wrestling. However, it is telling that the one major collection of wrestling criticism—Nicholas Sammond's edited collection *Steel Chair to the Head: The Pleasures and Pain of Professional Wrestling* (2005)—focuses on various dimensions of professional wrestling's relationship to identity politics (class, race, sexuality, gender, and the like) without thinking through how all of these relationships have a direct ground in its problematic relation to reality and violence. See Nicholas Sammond, ed, *Steel Chair to the Head: The Pleasures and Pain of Professional Wrestling* (Durham: Duke University Press, 2005).
13. Henry Jenkins, "'Never Trust a Snake': WWF Wrestling as Masculine Melodrama," in *Out of Bounds: Sports, Media, and the Politics of Identity*, eds. Aaron Baker and Todd Boyd (Bloomington: Indiana University Press, 1997), 50.
14. Ibid., 48.
15. Ibid., 63.
16. Ibid., 54.
17. John W. Campbell, "Professional Wrestling: Why the Bad Guy Wins," *Journal of American Culture* 19 (1996): 129.
18. Ibid., 127.
19. John Rickard, "'The Spectacle of Excess': The Emergence of Modern Professional Wrestling in the United States and Australia," *Journal of Popular Culture* 33 (1999): 133.
20. Ibid., 130.
21. Much of the following observations come from my own experience as a wrestling fan, that is, from my own memory of watching ECW in the late 1990s and early 2000s. However, one does have access to this footage through contemporary DVD collections, such as *The Rise and Fall of ECW* (2004), *Bloodsport: ECW's Most Violent Matches* (2006), and *ECW: Extreme Rules* (2007). These DVD releases are all the result of World Wrestling Entertainment's (WWE) purchase of the video library of Extreme Championship Wrestling in bankruptcy court ("WWE Launches ECW"). Terry Funk and Sabu's match, which I analyze later in the essay, is available on the *Bloodsport* collec-

tion. See "WWE Launches ECW as Third Brand," *2006 News*, World Wrestling Entertainment, last updated May 25, 2006: http://corporate.wwe.com/news/2006.

22. In his autobiography, Mick Foley remembers a televised interview in which he responds to a fan's hand-drawn sign that read, "Cane Dewey." ("Dewey" is Foley's actual son; the sign calls for another wrestler to attack him.) Foley includes the transcript of the interview in his book: "Dewey Foley is a three-year-old boy—you sick sons of bitches. . . . these letters behind me [ECW] are a blatant lie, and those fans who sit there and say, 'He's hardcore, he's hardcore, he's hardcore,' wouldn't piss on you if you were on fire." The intensity of the interview, in which Foley swerves in and out of being "Cactus Jack," troubles the parameters that normally structure the relation between fans and a wrestling villain. See Mick Foley, *Have a Nice Day!: A Tale of Blood and Sweatsocks* (New York: HarperCollins, 1999), 343.

23. Dalbir S. Sehmby, "Wrestling and Popular Culture," *CLCWeb: Comparative Literature and Culture* 4 (2002): http://docs.lib.purdue.edu/clcweb/vol4/iss1/5.

24. "History of ECW Championship," World Wrestling Entertainment, accessed June 23, 2010, http://www.wwe.com/shows/ecw/history/ecwchampionship.

25. Barthes, "The World of Wrestling," 20.

26. Baudrillard, "The Precession of Simulacra," 7.

27. Sharon Mazer, *Professional Wrestling: Sport and Spectacle* (Jackson: University Press of Mississippi, 1998), 10.

28. Barthes, "The World of Wrestling," 20.

29. Baudrillard, "The Precession of Simulacra," 1.

30. Jacques Derrida, *Of Grammatology*, trans. Gayatri Chakravorty Spivak (Baltimore: Johns Hopkins University Press, 1997), 167.

31. Marco Abel, *Violent Affect: Literature, Cinema, and Critique After Representation* (Lincoln: University of Nebraska Press, 2008), 20.

32. Roland Barthes, "The Reality Effect," in *The Rustle of Language*, trans. Richard Howard (Berkely: University of California Press, 1989), 146.

Chapter 14

Drafting the Hyperreal: Ownership, Agency, Responsibility in Fantasy Sports

Mihaela P Harper and Andrew J Ploeg

In just ten years, fantasy sports have grown from a statistics-obsessed subculture to an indisputable cultural phenomenon. According to a survey by the Fantasy Sports Trade Association (FSTA) released in 2008, nearly 30 million people in the United States and Canada play fantasy sports yearly, and the industry is estimated to have a "total market impact of $4.48 billion," $800 million of which is spent directly on fantasy sports products.[1] Already there are fantasy sports websites and magazines; advice shows on television, radio, and the Internet; statistics services; cell-phone applications; and even a situation-comedy (*The League*) about the lives of a group of people involved in a fantasy football league. Given this heightened popularity, it is virtually impossible to watch real[2] North American sports on television these days without being inundated by fantasy-specific updates, alerts, statistics, and analysis.

The ubiquity of fantasy sports, however, belies their humble origins. While the precise genesis of fantasy sports remains under some debate, their history and development can be most reliably traced through their longest-standing game, baseball. Though their progenitors—board and dice sports simulation games—predate them by a decade or two,[3] the *National Baseball Seminar*, invented in 1960 by sociologist William Gamson, is considered by many to be the "forerunner of [. . .] fantasy baseball."[4] Gamson's invention enabled participants to draft actual players onto imaginary teams and earn points based upon the on-field performance of those players within a specific set of statistics during the current baseball season. The *National Baseball Seminar* marks a watershed moment in the evolution of fantasy sports, in which they became not about simulating performance, based on a summary of actual past performance (as they were in board games), but rather about tracking and then translating the ebb and flow of player performance into fantasy points as they happen. Two decades later, Daniel Okrent, a writer and the first public editor of the *New York Times*, refined Gamson's game into modern-day "rotisserie baseball," an influential variety of fantasy baseball, named after the Manhattan restaurant where its inaugural season's participants held weekly meetings.

The meteoric rise in the popularity of fantasy sports owes as much if not more to the impact of the Internet, as it does to the development and refinement of the games themselves. Prior to the prevalence of the Internet, fantasy league commissioners had to track player statistics manually, painstakingly combing through daily or weekly box scores in order to calculate player and team performance and to generate overall league standings.[5] Today, given the widespread access to the Internet and the processing power of contemporary computers, statistics can be gathered and fantasy points can be calculated in real-time. Participants can now check their team's score, access player performance and health updates, adjust their lineups, scout free-agents, and/or make and accept trade offers based on up-to-the-second information at any time of the day or night. This convenience and immediacy has in large part fueled the interest in fantasy sports and broadened their appeal by reducing the investment of time necessary to participate in and to succeed at them.[6]

Despite superficial distinctions between modes of game-play—such as head-to-head and rotisserie[7]—fantasy sports are nevertheless united by their unique relationship to contemporary conceptions of the fantastic and the real, and, by extension, to the Baudrillardian notion of the *hyperreal*, which names the irremediable blurring of the two. In this paper, we argue that fantasy sports not only embody the hyperreal but, in the same gesture, also enable "owners" to participate in the fusion, not in order to preserve the reality principle but to partake in its obliteration. Our suggestion is that fantasy sports raise an awareness of the operative framework of the hyperreal (i.e., open the possibility for transforming dichotomous thinking) and, thus, simultaneously challenge Baudrillard's notion and make possible a kind of agency. This agency, which is not yet an action in the conventional sense, can rather be understood as a responsibility to thinking otherwise, to non-dichotomized thought.

In "The Precession of Simulacra," Baudrillard develops his notion of the hyperreal, a concept critical both to our argument for fantasy sports as both the hyperreal and a redefinition of it and to an understanding of cultural studies in the age of postmodernism. In his essay, Baudrillard fundamentally challenges the nature and even the existence of the real by subverting the idea that reality precedes representation, arguing instead that representation has no origin in the real. "It is no longer a question of imitation, nor duplication, nor even parody," he writes, "It is a question of substituting the signs of the real for the real, that is to say of an operation of deterring every real process via its operational double, a programmatic, metastable, perfectly descriptive machine that offers all the signs of the real and short-circuits all its vicissitudes."[8] According to Baudrillard, then, simulation is not based in reality precisely because there is no distinct reality left to base it in, while dissimulation, on the other hand, lies about the reality that is actually there. The latter keeps the principle of the real intact, while the former completely effaces it. He asserts that not only does representation precede reality itself but also that, by forcing what we conventionally think of as real to conform to the simulacra in postmodern culture, we have made the two indistinguishable. When simulation becomes at once representation and reality, we enter the world of the hyperreal (i.e., what we experience as reality when there is no reality to actually experience). In the hyperreal, the simulacrum is not the simulation of reality but its substitution and "murder"[9]—all distinction between the real and the imaginary has now vanished. Baudrillard, thus, paints a picture of a postmodern world without referent, in which sign and thing are indistinguishable and in which exit from its totalizing system—perpetuated, if not created, by media saturation—is preempted. Drawn on the basis of Marx's conception of commodity fetishism, hyperreality is not only an influential concept in postmodern philosophy and theory; it is also one that informs contemporary cultural and media studies.

Far from a pop-cultural phenomenon limited entirely to the realm of imagination, fantasy sports—a misleading moniker—embrace and actively participate in the constitution of the hyperreal through a competition in which owners draft real players from actual sports onto their own teams and manage them in an imaginary setting. Fantasy football owners, for example, make daily and weekly decisions regarding which players they will place in their starting lineups for that weekend's games (meaning whose statistics from actual NFL games will be used to determine the outcome of the fantasy contest), which previously non-rostered players to add to their team from the league's waiver wire, and, on occasion, for which players on opposing teams to trade. These decisions regarding actual players and imaginary teams form the basis of fantasy leagues and fundamentally blur the conventional distinctions between the real and the fantastic. They signal the emergence of hyperreality, and, as we contend, serve as a condition of possibility for both an awareness of the operation of the system and for critical engagement.

Cultural, social, and media theorists invested in Baudrillard's oeuvre disagree on the place of criticism and resistance within it, or, more specifically, whether there is space for a kind of agency in the hyperreal. Some arguments purport that there is no place, no position from which to resist, to act politically, to critique.[10] Such claims, as Michael L. Fitzhugh and William H. Leckie, Jr. suggest in "Agency, Postmodernism, and the Causes of Change," emerge due to an alignment of Baudrillard's premise with what is often perceived as the denial of the possibility for agency by postmodern theory.[11] More significantly, Baudrillard's own articulation of the world as hyperreal creates possibilities for reading his theory as fundamentally nihilistic, precisely because it appears to assert that the hyperreal, as Temenuga Trifonova writes, is "defined as the elimination of the subjective point of view, the suppression of the look, the fact that the object of perception is always already there, already seen, thus preventing the act of seeing."[12] A related assertion would be that the operation of the hyperreal appropriates agency by tricking consciousness into believing that it is in control of a system to which it is superior. To Baudrillard, this is the most significant and perfidious aspect of the hyperreal in that it pretends that it has what it does not have, since, in the hyperreal, reality and illusion, subject and object become indistinguishable and all acts are appropriated by the system as simulacra. Baudrillard puts it thus: "We are simulators, we are simulacra . . . we are concave mirrors radiated by the social, a radiation without a light source, power without origin, without distance, and it is in this tactical universe of the simulacrum that one will have to fight—*without hope*, hope is a weak value, *but in defiance and fascination*" (emphasis added).[13] We argue that what lies beyond the dash is a possibility for reconceptualizing the hyperreal, since, paradoxically, it bears a capacity for resistance and criticism, for a kind of agency within itself.

Baudrillard's exclusion of hope is not surprising, considering that, by definition, hope requires a distance and a telos, a distance that can be filled by the act of hoping, by hope itself, and an end, something that is hoped for, a dream or an illusion. Neither distance nor illusion, however, has a place in the hyperreal; it eradicates both. The inoperability of a recognizable hope is among the central reasons why Baudrillard's thought appears as bleak and unsettling as it does. But, in lieu of hope, something else emerges, defiance and fascination, both of which are already built into the system.[14] Baudrillard theorizes fascination in his chapter "On Nihilism" in *Simulacra and Simulation*, the same chapter in which he pronounces himself a nihilist because of his own complicity in the system and then brackets the pronouncement that nihilism, due to its teleology, is impossible. "Fascination," he writes, "(in contrast to seduction, which was attached to appearances, and to dialectical reason, which was attached to meaning) is a nihilistic passion par excellence, it is the passion proper to the mode of disappearance. We are fascinated by all forms of disappearance, of our disappearance."[15] Conceptualized in opposition to seduction, fascination removes the dichotomy of real and imaginary, which is the condition of possibility for meaning and, thus, for critique. And since, to Baudrillard, the elimination of the dialectical is tantamount to emptying the critical, fascination in the hyperreal appears to operate only as an uncritical self-perpetuation of the system. What makes this conception problematic, however, are precisely the affinities between fascination and fight, fascination and nihilism, which Baudrillard articulates not as possibilities for a reversibility in the sense of a return to a prior order, but rather as the capacity of the hyperreal to relate to itself. Along this line, our suggestion is that fascination (of the hyperreal with itself, which is, in the same sense, ours with the hyperreal) makes possible a certain kind of awareness, a defiance that is agential in its very fidelity to the non-dichotomous. Thus, while neither hope nor agency in the conventional senses (and binary philosophemes) are recognizable as such in Baudrillard's hyperreal, fantasy sports illuminate the possibility of fascination to operate as a response of the system to itself, as a kind of agency in terms of critical awareness.

In a rudimentary sense, fantasy sports appear as a (techno)logical extension of a "passion" for sports. Okrent, in his introduction to *Rotisserie League Baseball*, explains the impetus behind the invention as "[the raising] of the ineluctable movements of generations enthralled by baseball to another, higher plateau."[16] Translated into the framework of the hyperreal, Okrent's articulation reveals fantasy sports as an embodiment of the fascination theorized by Baudrillard, a passion not only of the hyperreal for its hyperreality but also a fascination appropriated by the system through us. Implicit in the explanation, however, is also a quest for a different kind of engagement, driven by the seemingly inherent perpetuation of fandom. Within the framework of this implication, fantasy sports appear as a solution to the problem of the interest of succeeding generations in new modes of engagement. In a variety of significant ways, however, fantasy sports not only do *not* present a solution, but instead pose a challenge by creating problems, conceptual, juridical, and internal to the system of fandom, sports, and the hyperreal.[17] They challenge the very "passion" and fascination that drives them by forcing an interrogation of the premises upon which both of these operate. The designation of fans that do not engage in fantasy sports as "purists" attests to this interrogation. "Purists," an unofficial label reserved for fans who subscribe to a "traditional" conception of fandom and team allegiance, deny a reality to the expression of a passion for sports or of fandom as such in the form of fantasy sports. From the "purist" perspective, fantasy sports are literally fantastic, unreal, and, since the existence of the unreal depends on its opposition to the real, then fantasy sports, by nature, reinforce the reality principle. This logic is one of the remarkable ruses of the system, and, as we suggest, one of the central reasons for the critical capacity both of Baudrillard's theory and of fantasy sports as the hyperreal.

Despite the "purist" sentiment, fantasy sports have grown so popular and participant investment in them so intense that they have triggered a reconceptualization of what it means to be a "fan" in today's postmodern, media-saturated culture. In his essay on the "reality" of fantasy football, Michael Serazio writes that, when football fans become fantasy football owners, "the game changes shape for the individual, even if, in reality, nothing has changed at all on the field of play. We seem to watch the game differently. We relate to it in new ways."[18] He elaborates this point, exposing several relevant implications:

> Fantasy football reduces teams to individual performances. It also takes fans who previously formed communities based upon allegiances to those teams and reduces them to individuals rooting for patchwork lineups. [. . .] The very notion of a fantasy football "team" is obviously a fabrication. [. . .] But it is a fabrication that [. . .] represents a fitting sports extension of the themes apparent in an individual-centered, personally-tailored media era.[19]

Serazio argues that fantasy sports are changing fandom and, along with it, the way sports are experienced. The fan, he contends, is no longer simply a passive observer of sports; rather, through the illusion facilitated by fantasy sports, she is an active participant who "somehow shares in the action and the glory of the players he 'owns.'"[20] This conceptualization of fantasy sports as "illusion" and "fabrication," however, falls prey to a logic that reinforces the reality principle by implying an essential dichotomy of truth and falsity, reality and fantasy, a logic fundamentally at odds with the Baudrillardian hyperreal. The polarity, implied in Serazio's essay and common to the increasing but still limited amount of fantasy sports scholarship, is replaced by undecidability in the hyperreality of fantasy sports. Reality is no longer diametrically opposed to fantasy; it is indistinguishable from it. Ironically, in *fantasy* sports, fans abandon the dichotomized framework of fantasy/reality and, instead, wholly embrace the paradigm of the hyperreal.

On a practical level, the operative framework of fantasy sports is perhaps most clearly exemplified by the high percentage of fans who are also fantasy sports owners (i.e., "fan-tasy owners"), for whom the lines between actual and fantasy sports are blurred to the point that many often hold success in either realm (realms that can be and are, at times, at odds) in the same regard. It is not simply a matter of inverting the hegemony of reality over fantasy but of blurring the two to the point of indistinction. In fact, it is not uncommon, and is for the most part unavoidable, for "fan-tasy owners": a) to cheer for their favorite real team, only if the team's success is due to the positive performance of a player who is also on their fantasy team; b) to be torn between a favorite team's victory and the success of a player on both the opposing team and their fantasy team, or c) to watch a game that they have no direct real-world interest in, but to pull for players on competing teams who are also on their fantasy team. All three of these scenarios involve interests that would normally be at cross-purposes; yet, in the logic born from the hyperreality of fantasy sports, they align perfectly. In these situations, it becomes clear that there is no longer a distinction between or hierarchy of real and fantasy; thus, many "fan-tasy owners" place upon a fantasy victory as much urgency and importance as they do on a real one. Given the investment of time and effort, as well as the stakes at play in fantasy sports (i.e., sports knowledge, prognostication ability, bragging rights, trophies, money, etc), it is not difficult to track and even sympathize with the seemingly capricious and elusive loyalties discussed above. Balancing them, however, indisputably changes the role, responsibility, and ethics of the fan, who now bleeds a hyperreal fusion of his real team's colors and those of his fantasy team.

The apparently seamless meshing of the real and the fantastic, in these scenarios, makes fantasy sports yet another example of Baudrillard's hyperreal. This view can be complicated, though not contradicted, by the assertion that there remains a subtle and internal privileging of the real at work in fantasy sports. In other words, it could be argued that fantasy sports are contingent upon the reality of sports in order to exist—if there were no real sports, there would be no fantasy sports, while the reverse could not be the case. This assertion bears a certain validity: without statistics, derived from the quantifiable feats of actual athletes, fantasy sports could not be played, or they would be forced to generate their own statistics through game-play simulation (a reversion to their board game roots) or through video-game-style player production (a move that would challenge their very status as hyperreal by moving them entirely into the register of fantasy). However, it is precisely the reality of real sports that is fundamentally in question in this assertion and is, in fact, interrogated by the very framework of fantasy sports. In a Baudrillardian sense, this reality is, itself, already a thoroughly hyperreal interpenetration of politics, popular culture, money, titillation, and performativity. Sports are, in essence, performance, not simply on the level of athletes performing sporting feats for the entertainment of an audience but also on the level of sports reproducing themselves, performing what sports should or could be like. They are entirely mediated and, thus, not as much sports as the image of sports—an image of an image, which always precedes the real that is no more. Baudrillard writes, "Simulation is no longer that of a territory, a referential being, or a substance. It is the generation by models of a real without origin or reality: a hyperreal."[21] Players on the field are no longer simply players, but models, inseparable from their marketed personas, their endorsements, their fame, their fashion, and their riches. They are simulations, grounded in nothing but an image that is itself already an image. Thus, fantasy sports, while driven by statistics derived from actual sports and real players, are not based on a reality. They are, instead, a hyperreality, resting on a reality that is itself also a hyperreality, an endlessly circular and repeatable system, which seemingly leaves no room for agency and renders resistance impossible. We contend, however, that in this perpetual circularity a different gesture is born, one that reveals the quest of the system to relate to itself and to locate an exit from itself.

The way in which fantasy sports question the conventional dichotomy of reality and fantasy suggests that it is not only the fan that has changed but also the relation of the fan to her fantasy. This relation is in flux, in that it is no longer tantamount and confined to a relation with a totality (i.e., an entire team, geographic location or particular game, identity or set of colors). It is redefined, exploded, or, to draw on Jacques Derrida, *exappropriated* as multiplicitous, reversible, ultimately unstable, which is precisely what allows an awareness of the operation of the hyperreal to emerge, an awareness of non-dichotomized thinking that is itself a kind of agency. Fantasy sports make visible the possibility for a kind of exappropriation, for the system no longer to chase the specter of the real that haunts it but, instead, to defy the hyperreal in the same gesture of self-relation by which it seeks to perpetuate itself. In other words, it disowns itself in the movement in which it appropriates itself, a movement that Peggy Kamuf captures compellingly:

> The double prefix "exap-" marks the sense of "-propriation" with an irreducible discordance or dissociation between its two directions. Whereas the proper movement of the proper can only be in an *ap*propriative direction back to itself, the circle of return cannot complete itself without also tracking the contrary movement of *ex*propriation. The more it seeks to keep itself to itself—uncontaminated, purely proper—the more "propriation" loses itself in the "ex-" of an exteriority to itself. The more it seeks to remain faithful to itself, the more it betrays itself.[22]

Similarly, the more the hyperreal, fascinated with its own hyperreality, seeks to respond to that fascination and to identify itself in relation to a perfidious fantasy—reality dichotomy, the more it betrays itself in a movement of dis-identification. In other words, fantasy sports, while arising from within the system in order to embody it, create the possibility for exappropriation, awareness, and, thus, for agency. Fantasy team owners, for instance, are compelled to defy the traditional conceptions under which their engagements with sports had operated, and, in some sense, to dis-identify themselves. Thus, fantasy sports open the system not to a space for a dialectical, critical position proper (i.e., a traditional, totality-invested fandom) but for a multiplicitous movement of positions (i.e., a mobile plurality of team and player engagements), or rather an awareness of possibilities not dependent on a reality-fantasy divide.

Although this awareness is not immediately recognizable as agency, because it is not legible as conventional action, we propose that it bears in itself both defiance as fight and, significantly, a kind of responsibility to respond to the system as it responds to itself. In *The Gift of Death*, Derrida points out that responsibility (and, we would like to add, agency) has historically depended on action, a "doing, a *praxis*, a *decision* that exceeds simple conscience or simple theoretical understanding . . . [that] requires a decision or responsible action to answer for itself *consciously*, that is with knowledge of a thematics of what is done."[23] He interrogates the conventional conception of responsibility in order to point out that divorcing it from action opens it to difference, to a responsibility that is not ready-made, but involves instead new forms of engagement. In the framework of the hyperreal, the very impossibility of making a distinction between real and fantasy opens possibilities for such new engagements, and, we argue, fantasy sports give rise to an awareness of multiplicities that is itself this kind of responsibility.

In everyday discourse, Derrida contends, "one treats as nihilist, relativist, even poststructuralist, and worse still deconstructionist, all those who remain concerned in the face of such a display of good conscience."[24] The "good conscience" referenced is precisely the subscription to thinking the world in dichotomous terms, rejecting an awareness of what is already its capacity for reversibility (in the Baudrillardian sense), its irreducibility to terms, such as fantasy and reality. Fantasy sports exhibit the infinite possibilities of the hyperreal in terms of dis-identification as opening and changing not only the ways in which we perceive our allegiance to and place in the system but also the relation of the system to itself. This kind of agential awareness, we suggest, is the possibility that both Baudrillard and Derrida (despite their differences) see for a struggle, and the latter, in particular, finds crucial to a necessarily ever-changing engagement with teletechnologies in order to be able to "to respond, select, intercept, intervene"[25] in the system and in our own lives.

To announce, as Baudrillard does, that the hyperreal is "total simulation, without origin, immanent, without a past, without a future, a diffusion of all coordinates"[26] is not to announce a catastrophic end. Rather, it is to indicate the emergence of a different register, a different mode of operation that, nevertheless, necessitates critical engagement, radicalization, opening. This is a "must" that reverberates throughout Baudrillard's works, as well as the oeuvres of other postmodern and poststructuralist theorists, and infuses his conceptions with a certain ethics. As Derrida and Stiegler note, "we must learn, precisely, how to discriminate, compose, edit"[27] in terms of technology. In such engagements lies a response to the system and to ourselves, which is also a responsibility to creating *new forms of intelligibility*; and, it is in such engagements that our awareness and agency express themselves. In this sense, we contend, fantasy sports operate as a new form of intelligibility, one that makes possible new loyalties and responsibilities, dissociated from conventional dichotomous thinking. They facilitate new ways of perceiving and relating to ourselves and the system of which we are a part, an awareness of the hyperreal that is itself a kind of agency and responsibility and that enables the infinite possibilities of non-dichotomous thought.

Notes

1. "Fantasy Sports Industry Grows to a $800 Million Industry With 29.9 Million Players," *PRWeb*, July 10, 2008, accessed June 15, 2010, http://www.prweb.com/releases/2008/07/prweb1084994.htm.

2. Although we draw on Baudrillard's contention that terms, such as "real," "fantastic," "actual," "imaginary," etc., have been rendered defunct by the hyperreal, we are, nevertheless, impelled to use them in their conventional senses in order to explicate the ways in which fantasy sports are commonly thought to operate, namely through the reality-fantasy dichotomy.

3. Most notably *All-Star Baseball* (1941) and *Strat-o-matic Baseball* (1954).

4. Kevin G. Quinn, *Sports and Their Fans: The History, Economics and Culture of the Relationship Between Spectator and Sport* (Jefferson, NC: McFarland, 2009), 141.

5. Ibid.

6. Michael Serazio, "Virtual Sports Consumption, Authentic Brotherhood: The Reality of Fantasy Football," in *Sports Mania: Essays on Fandom and the Media in the 21st Century*, ed. Lawrence W. Hugenberg, et al. (Jefferson, NC: McFarland, 2008), 230.

7. In head-to-head leagues, a single team faces only one other team in any given week, and the team with the most points by the end of that period is the winner. In rotisserie leagues, on the other hand, each team competes against all other teams in the league at the same time by accumulating statistics in a given set of categories over the course of an entire season.

8. Jean Baudrillard, *Simulacra and Simulation*, trans. Sheila Faria Glaser (Ann Arbor, MI: University of Michigan Press, 1994), 2.

9. Ibid., 8.

10. Jaber F. Gubrium and James A. Holstein's definition of agency through a "capacity to effectively act upon the world" belies a number of these arguments; our work seeks to interrogate and redefine this sense. Jaber F. Gubrium and James A. Holstein, "Individual Agency, the Ordinary, and Postmodern Life," *The Sociological Quarterly* 36, no. 3 (1995): 555.

11. Michael L. Fitzhugh and William H. Leckie, "Agency, Postmdernism, and the Causes of Change," *History and Theory* 40, no. 4 (2001): 59-81. Similarly to Baudrillard's project, the works of other contemporary thinkers have been accused of nihilism; however, close readings of those works reveal an interest in opening productive possibilities for kinds of agency: subjectivization (Foucault), radical opening (Derrida), and becoming (Deleuze and Guattari) to mention just a few.

12. Temenuga Trifonova, "Is There a Subject in Hyperreality?" *Postmodern Culture* 13, no. 3 (2003): sec. 20.

13. Baudrillard, *Simulacra*, 152.

14. This point invokes two different assertions: Baudrillard's that resistance to the control a system exercises is inherent, and Michel Foucault's that lines of resistance inhere in and even precede relations of power.

15. Baudrillard, *Simulacra*, 160.

16. Daniel Okrent, "An Introduction to Rotisserie League Baseball," in *Rotisserie League Baseball*, ed. Glen Waggoner (New York: Bantam, 1984), 4.

17. Fantasy sports interrogate not only the reality-fantasy dichotomy, but have also required a rethinking of the definitions of sports gambling and the rights to the licensing of player names and statistics.

18. Serazio, "Virtual," 239.

19. Ibid., 241.

20. Ibid., 229.

21. Baudrillard, *Simulacra*, 1.

22. Peggy Kamuf, ed., "Introduction: Reading Between the Blinds," in *A Derrida Reader: Between the Blinds* (New York: Columbia University Press, 1991), xxviii.

23. Jacques Derrida, *The Gift of Death*, trans. David Wills (Chicago: University of Chicago Press, 1995), 25.

24. Ibid., 85.

25. Jacques Derrida and Bernard Stiegler, *Echographies of Television*, trans. Jennifer Bajorek (Oxford: Polity, 2002), 69.
26. Baudrillard, *Simulacra*, 125.
27. Derrida and Stiegler, *Echographies*, 59.

Part IV

Hyperreal Homes: Culture and Identity in Postmodern Places

Chapter 15
Room(s) for Sight: The Artificial Everyday Theater of Rimini Protokoll

Matt Cornish

About a year ago, in need of confirmation of an Amazon.com order, I sat in front of my computer, slightly confused, waiting for my cell phone to ring. I had just entered my phone number into their website and sure enough, in less than a minute, I was speaking to a call center operator in India. Just two or three minutes later, we sorted out the situation, and my order of boxed Indian food was on its way to Connecticut. In an increasingly "connected" world, we are by no means unaccustomed to Internet-enabled communication, whether through computers, or cell phones, or some mixture of the two. That's why, when I entered a room alone at the Goethe-Institut New York, sat down on a couch, and waited for a cell phone sitting on a table to ring, I did not feel that anything was out of the ordinary. But instead of speaking briefly with an anonymous Indian call center operator about a purchase, I had a genial one hour conversation with Subhaditya Das of Descon Limited in Kolkata. I was participating in the "intercontinental phone play" *Call Cutta in a Box*, created by German theater provocateurs Rimini Protokoll.

Founded in 2002 by Daniel Wetzel, Helgard Haug, and Stefan Kaegi, all graduates of the Gießen Institut for Applied Theater Studies, Rimini Protokoll strives to shatter the conventions of the major state funded institutions of Europe, which produce new and classical plays in proscenium theaters. Like the similar American groups Nature Theater of Oklahoma and Radiohole, Wetzel, Haug, and Kaegi belong to a generation raised on cable television and the beginnings of the Internet, and show a deep interest in exploring the interplay of representation, performance, and the everyday existence we call reality. Old enough to remember the Berlin Wall, but young enough to have come of age artistically after it came down, Rimini Protkoll chronicles the effects of global capital in a post-Cold War world.

Working together as directors, dramaturgs, and adaptors, in a threesome and just as often in pairs or by themselves, Wetzel, Haug, and Kaegi frequently create productions starring "experts." These individuals may have real-life expertise as, for example, soldiers or elderly nursing-home patients (they are experts on their biographies), but have no experience as performers. Rimini Protokoll generates the content for their productions from their experts' lives and stories—antecedents to which can be found in documentary theater, influentially pioneered by German directors Erwin Piscator and Bertolt Brecht in the 1920s and 30s. The theater-works the group creates examine the intersections of these non-mediated performers (amateurs on the stage) with contemporary technology (Skype, video projections), as well as public History with private memory, and national with global identity. They destabilize spectators' relationships with these binaries, sometimes by engaging them as participants, and sometimes by exposing them to members of their society they overlook—and occasionally both.

In *Deutschland 2* (Theater der Welt Festival, 2002) for example, Rimini Protokoll cast 237 residents of Bonn as parliamentary representatives; listening to a live feed of a debate in the German parliament, the participants created a simultaneous copy of it. For *Torero Portero* (Goethe Institute Córdoba, 2001), Kaegi engaged three Argentinean porters to recount their lives, standing in the street as spectators watched from a distance. The production originated in Córdoba, and toured to Europe and later Brazil—where local Brazilian porters were added to the cast. Two productions in particular—*Wallenstein* (Internationale Schillertage Mannheim, 2005)[1] and *Call Cutta in a Box* (multiple locations, 2008)[2]—speak to Rimini Protokoll's interest in exploring how we construct meaning and "reality" in an increasingly mediated, globalized, and commercial everyday that resists the frames and reliability we seek.

In *Wallenstein*, subtitled a "documentary production," Rimini Protokoll put on stage several individuals, none of them professional performers, to tell stories from their lives, including: Dr. Sven-Joachim Otto, a Mannheim politician and judge; Esther Potter, an astrologer; Dave Blalock and Darnell Stephen Summers, American Vietnam veterans and anti-war activists; Friedemann Gassner, an electrician who also memorizes Friedrich Schiller's texts; and Ralf Kirsten, a former police commissioner of the German Democratic Republic. These "experts," from Weimar and Mannheim (on opposite sides of the iron curtain during the Cold War), recount bits and pieces of their biographies in a format loosely connected to Schiller's trilogy *Wallenstein* (1799), a historical drama about the downfall of a German general during the Thirty Years' War.

Wallenstein takes place on a stage open to the audience, who are seated on risers directly in front of the action, with a turn-table center stage and a movable metal wall stage-right. The space is eminently adjustable and simple, with the only accoutrements being colored tape on the floor, campaign posters on the wall, and a video screen stage-left. The performance begins when Friedemann Gassner, the electrician and Schiller fan, walks on stage in a red soccer-referee shirt, black shorts, a whistle around his neck, and what looks like a yellow penalty card sticking out of his shirt-pocket. He quotes Goethe on Schiller, and pulls out a pocket copy of *Wallenstein* (what had appeared to be the yellow card) before proceeding to memorize a new verse from the play. Gassner organizes the remainder of the two-hour production by blowing his whistle to signal scene changes; Rimini Protokoll divides the evening into four parts, along the lines of Schiller's trilogy, with an introduction and sections, announced out loud, devoted to each of the three parts of *Wallenstein*. The group divides the experts' stories so that they unfold throughout the performance, and coincide roughly with the dramaturgy and themes of Schiller's play (for example, Dr. Sven-Joachim Otto, the Wallenstein figure, does not enter until the second section, and his political ruin takes place at the end of the production, echoing Schiller).

With their *Wallenstein*, Rimini Protokoll creates a new history play for the post-1989 German nation. Questioning the status of text as primary aesthetic signifier, the establisher of meaning and historical facts, Rimini Protokoll emphasizes individual oral story-telling (as well as far more modest characters, politically). The dramaturgy of Rimini Protokoll's *Wallenstein*, with interruptions from (supposedly) unplanned cell-phone calls and interspaced with projections of images and video, does not allow for the proto-Hegelian progress created in Schiller's monumental play. Dialectical History, whether the communist march of Marx or Francis Fukuyama's democratic-capitalist End of History, advances through disasters and catastrophes (as Frederic Jameson writes, paraphrasing Henri Lefebvre, in *Postmodernism*[3]). And, as Walter Benjamin writes in *Theses on the Philosophy of History*, "There is no document of civilization which is not at the same time a document of barbarism."[4]

Rimini Protokoll provides a possible alternative model in their *Wallenstein*. The text of Schiller's play appears throughout the performance, always in the yellow pocket edition, but does not take control of the evening, as the actors use it as a prop to stand in for land-mines or grenades (for example), instead of as authoritarian guide. In addition to the spoken stories, Rimini Protokoll shows documents, like photographs of the veterans as young men in Vietnam, as supporting material, proof of the stories' authenticity, allowing the audience to invest emotionally. The personal memories that supersede a written text, re-embodied on the stage and repeated (in each performance) for a public audience, move through remembrance towards reconciliation: a kind of talk therapy, not dissimilar from the proliferation of memoirs and solo-performances.

But Rimini Protokoll does not create an alternative "authentic" history. Rather than privileging individual oral memory over dramatic text, this *Wallenstein* instead reveals its own status as a simulation, a construction. The "unplanned" phone calls—from men to Rita Mischereit on her cell-phone, in hopes of utilizing her agency to organize extramarital affairs—come adorned with the "signs" of improvisation: Mischereit apologizes to the other "experts" for interrupting their stories, they shush the audience so as not to alert the callers. On the stage however, with cues provided by the whistle-totting Gassner, not to mention the light-board operator and other behind the scenes crew members, the calls become questionably "real." The audience members cannot tell, and know that they cannot tell, if the phone calls are part of the artifice—or, if they in fact are *interruptions* of the simulation, the performance.[5]

This ambiguity extends to the "expert" participants themselves. The motivations of the individuals telling their stories become increasingly unclear as the performance moves forward: despite his engaging and funny back-room political war stories, Dr. Sven-Joachim Otto cannot be trusted; he could just be continuing to spin facts for future political gain. He calls attention to this at the end of the performance. As the walls of the space close in around him, he tells of betrayal by members of his own party, and then asks the audience members if they would vote for him now, and if they feel "sympathy" for him. Many raise their hands in answer to the second query. The moment highlights questions that have been lingering anyway: Why did these individuals on stage answer Rimini Protokoll's casting call? What do they have to gain? On the stage, every adult—amateurs as well as professional performers—will become self-conscious and will "perform," if only themselves; in other words, they dissemble. In *Simulacra and Simulation*, Baudrillard argues that our culture necessarily destroys the mummified Ramses by forcing his body into the light to be inspected and displayed—"transplanted from a slow order of the symbolic, master over putrefaction and death, to an order of history, science, and museums.[6] Only through this process does the mummy become "real" to us (by becoming a simulacrum, the original decayed and revived with science[7]). Rimini Protokoll raises similar questions in *Wallenstein*: does the public display of secret memories, through which they become manifest ("real") to the audience, destroy the very "authenticity" audiences crave?

Baudrillard condemns more than the artificial object the attempt by society to *hide* the object's artificiality (citing as an example the repatriation in New York of the cloister of Saint-Michel de Cuxa[8]). Many of the artistic precedents for non-professional performers rely on the performers' un-tuned bodies and voices and their distance from the lies of the theater to give the artists' messages greater emotional or intellectual impact (the American group The Living Theatre provides just one example of this). Rimini Protokoll does the opposite here, not attempting to cover up the artificial with a mask of the original. *Wallenstein* relies on artifice that the audience can read *as artifice* for public commemoration. This becomes strikingly apparent when in the middle section of the performance, titled "The Piccolomini" after the second part of Schiller's trilogy, Gassner discusses how a particular piece of poetry by Schiller saved him from a deep depression when his lover walked out of his life. ("You had luck with this verse," Mischereit mentions, "because *Wallenstein* is actually quite boring. "*Wallenstein* is bleak," Gassner agrees, much to the delight of the audience.[9]) Like Polixenes in Shakespeare's *The Winter's Tale*—"This is an art / Which does mend nature"[10]—Gassner recognizes that the lies of art do not hide or disfigure the truth, but can reveal and restore it. Importantly, Gassner does not sit at home and silently sip his Schiller alone on the coach, but joins a Schiller-Bund (a Schiller club), where he can learn from and share with others. And he speaks the words out loud, embodying them by memorizing verses as he walks around his apartment—he performs this in the beginning of the play, working on the lines of the watchman from the second act of Schiller's *Wallenstein.*

But Gassner stands on stage with Hagen Reich, formerly a German soldier enrolled in army officer training school. Dressed in camouflage and helmet, Reich walks the audience through the training program, demonstrating mine-detection and other skills necessary for soldiers preparing to deploy to the Bosnian Civil War as members of NATO (as he was at the time). In the final section of the performance ("Wallenstein's Death"), Reich vividly describes how in one scenario, he was leading a group through a "mined" area, when he saw a violent confrontation at a "wedding" on one of his video-monitors. Losing his ability to distinguish between the scenario and the "real," Reich crossed the mined area to stop the violence at the wedding. He failed out of the program, and says that he has been unemployed ever since. Clearly shaken in remembering the incident, as well as the other training exercises (in one, he takes special note of the "real sausages" that were being cooked during the scenario), Reich seems to be experiencing a form of post-traumatic stress disorder. Not from an *actual* war experience, but from a simulated war—a particularly stimulating, violent, stressful theme park. The audience must react to the performance of these memories—utterly removed from the experience of walking through a field in the former Yugoslavia searching for mines.

I should note that I did not attend a live performance of Rimini Protokoll's *Wallenstein.* Instead, I watched a film of the production made for a television broadcast: a simulation. But how does video-simulation work with *Wallenstein*? Video of this live performance presents several obstacles to analysis, including: the entire stage was not visible to me at all times; I could see the actors' faces in close-up, allowing me to notice details like microphones that may or may not have been evident to the live audience; and I could rewind and rewatch scenes at my desire, breaking the continuity of the performance.

Some theater scholars, most notably Peggy Phelan, argue that there is a clear ontological distinction between live performance and video. "Performance's only life is in the present," she writes in *Unmarked: The Politics of Performance*. "Performance cannot be saved, recorded, documented, or otherwise participate in the circulation of representation *of* representations: once it does so, it becomes something other than performance."[11]

Philip Auslander strongly challenges these ideas in *Liveness: Performance in a Mediatized Culture*, arguing "that the relationship between live and mediatized forms and the meaning of liveness be understood as historical and contingent rather than determined by immutable differences."[12] I would agree with Auslander that critiques like Phelan's fetishize "theater" and the live body, creating reductive binaries. In viewing a video of a live performance, we are not engaging with something "other than performance," as Phelan says.

With the video, and assisted by photographs and critical reviews to help make up for the biases of the camera-operators and editors, we can think ourselves into the seats, and provide thorough and sophisticated analyses of most events—analyses that, unbeholden to hastily scribbled notes and dissolving memories, may actually be assisted by the ability to rewind and review, especially if the performance took place primarily in a language foreign to the critic. When acknowledged, there should be no fear here of discarding the experience for the copy, or of mistaking the one for the other; a video of a live performance is a record that stands *in between* the two, a live event no longer live, allowing students and critics to imagine the past event, if never re-experience it.

Watching a video of *Wallenstein*, my heightened awareness of media and representation encouraged sensitivity to those themes in the production itself. Rimini Protokoll thrives in gray areas like the one here between live performance and a video of it; they create metaphorically and often physically unenclosed spaces in which the audience must decide the border between real and fake, between restorative artifice and destructive s(t)imulation. Apart from the rather conventional use of theater space in *Wallenstein* and other productions, Rimini Protokoll has created many live events completely removed from the walls of theater buildings, stretching and sometimes breaking the famous definition by Peter Brook (in *The Empty Space*) of theater as at least one performer and one spectator sharing a room. In 2005, the team combined the aesthetics of radio play, city tour, and globalized economics to create *Call Cutta*, in which a call center employee in Kolkata led individuals on a tour of Berlin over a cell phone, exploring history through geography along with globalization and outsourcing. Playing with the themes and signs of theater, at one point the call center employee asked the participant to stand in the middle of a courtyard and sing for anybody who might be in an apartment looking down. For *Call Cutta* participants, Rimini Protokoll made the global feel local, the local feel global, and everyday reality feel performed.

In 2008, Rimini Protokoll took *Call Cutta* on the road, renaming it *Call Cutta in a Box* and tailoring the experience to individual cities, spaces, and people. Rather than leading audience members on a tour of a city, *Call Cutta in a Box* unravels the mysteries of a single room and two people connected by phones and satellites, forging sixty minutes of conversation and discovery (of talking and listening) out of the busy lives of consumers and laborers in capitalist behemoths like Manhattan and Kolkata. In early 2009, I traveled to the Upper East Side of New York City to participate in *Call Cutta in a Box*.

After being assigned to a space by an employee in the lobby of the Goethe-Institut New York, I walked alone into a sparsely furnished room with regal wood-paneling, and waited for the cell phone on the coffee table to sound. Once the phone rang, connecting me to Subhaditya Das in Kolkata, *Call Cutta in a Box* began with initial awkward questions (as when Das asked me: "Can you pronounce my name?"). This gave way to increasing depth as we got to know each other, chatting about religion, reincarnation, and travel experiences. Das carefully led me through pre-planned talking points and events, while allowing us some freedom to find our own conversation within the readily apparent superstructure created by Rimini Protokoll. He brewed me a cup of tea (the operators can control every element of the room remotely), sang to me in Bengali while asking me to imagine a related scene, and then told me to stand on a raised platform ("Does it remind you of a stage?"). Towards the end of the hour, he turned on the computer, showed me a map of Kolkata, printed out pictures of his life in India, and even requested that I perform a song for his amusement and edification. In a magnificent *coup de* "mediatized" *théâtre*, towards the end of the hour Das took his webcam and pointed it at a different computer; on that screen, I saw another participant in *Call Cutta in a Box* (whom I had met in the Goethe-Institut lobby prior to entering my room) talking and waving. Separated by satellites and cable lines, and layers of monitors and digital cameras, that other person was actually a mere twenty feet away, in the next room, performing in and watching the same play as me. It's hilarious, and tragic. Finally, we had to say our sad goodbyes, and Das let me know that I could email him anytime. (I did in fact email him about two weeks after our conversation; regrettably, he didn't respond).

By giving the subjects of their works voices, Rimini Protokoll intends to avoid the anxiety of mimesis in tackling post-colonial subjects. There is a long history of Orientalism in the theater, of projecting an "authenticity" that emerges from pre-modern or primitive "innocence" upon characters representing the non-European Other, while universalizing this Other and destroying the heterogeneity of differing cultures and individuals. This happens perhaps most damningly in the early twentieth-century avant-gardes like Expressionism, as well as with artists like Antonin Artaud and Bertolt Brecht. For Rimini Protokoll to simply *represent* the call-center operators (with paid actors), instead of using Descon employees, would enact, as Gayatri Spivak writes in "Can the Subaltern Speak?" "epistemic violence"[13] upon the very people they seek to connect with. Despite avoiding problems of representation, as artists they must inevitably structure their works—and certainly do so here—taking into their own hands the single most powerful aspect of art: the frame. In *Call Cutta in a Box*, Rimini Protokoll explores these tensions between artists and subjects, Europe and post-colonial India, representation and what we experience as life.

To begin with, *Call Cutta in a Box* reveals the utter strangeness of our new globalized economy. The idea of phoning a call center to ask about a flight from Hartford to Los Angeles and having the problem resolved by somebody in India, who may never have been more than few hours away from his or her home, isn't just ridiculous, it's absurd. By estranging the everyday experience of speaking with a call center operator, while embracing technology to bring two distant people "together," *Call Cutta in a Box* briefly makes the hugeness of globalization tangible and understandable. Technology becomes the hero in this story, allowing participants to see anew the radical changes global capitalism has wrought in our everyday lives. The room becomes a microcosm, a globalized cabinet of curiosities. Rimini Protokoll puns on the German word *Wirtschaftsraum*: a market or economic area, literarily translating as "economy-room." The room at the Goethe-Institut holds key objects in the creation of a "flat" world: computer, cell phone, printer, and stereo, all plugged in to the Internet, and thus controllable from almost anywhere. Some hidden gifts make New York feel even closer to Kolkata—especially the pan masala (given at the end), a sensual reminder of life in India, which, because it involves taste and smell, feels much more verifiable, more *real* than looking at pictures or speaking over a cell phone. By estranging participants from the operator while simultaneously making them feel closer, Rimini Protokoll finds the specific, the *human*, in the unfathomable hugeness of global capitalism. Out of dialogue, one might say, emerges the opportunity for the participant *and* the call-center operator to recognize the other's self-consciousness. And in fact, I did become familiar with Das. He shared with me his dreams of travel, his commute between home and work, the culture that helps him to identify himself. Naturally, I did the same.

In this *Wirtschaftsraum* consumers and suppliers mimic the service economy in miniature. My personal "expert" in the workings of global capitalism, Das printed out a picture for me close to the end of the hour. "Do you know who this is," he asked. The image of a middle-aged woman, smiling while standing in a kitchen and cleaning a jar, emerged from the printer. "Is it your mother?" I answered. "No, no," he said, "that's my servant." In this jarring moment, I realized that my operator is just a twenty-first century servant, doing his job, making me as comfortable as possible. (And in fact, the operators can do this literally, as they can change the temperature of the room.) The power dynamic in our relationship was never as equal as it initially felt. *Call Cutta in a Box* never allows you to forget that the moments of intimacy and trust are manufactured and artificial: a product created by Rimini Protokoll and delivered by the operator, an experience you bought when you purchased your ticket. The call center employee brews you tea, gives you little presents, operates all the electronics for you, answers your questions, and carefully tracks the hour you have together until the final seconds tick off. You are not just talking to your operator. He is working for you, and Rimini Protokoll makes the operator visible as a laborer.

In fact, by speaking English, using colloquialisms, and talking about the weather in New York, call center employees (whether working for Descon selling I.T. products or as a part of *Call Cutta*) *simulate* closeness, papering over otherness more than revealing it. India, which decided to teach English as a second language in the wake of British rule, is the world's largest service economy, and Rimini Protokoll uses the legacies of colonialism in that service economy to their advantage. They make evident the means of production in this new master-servant economy: the banter and quiet assistance that make the "other" more familiar, the experience of talking with an Indian call center employee more comfortable. In this way, *Call Cutta in a Box* is more about discovering the laborer as laborer than the laborer as person.

Indeed, instead of becoming truly familiar with Das, I became increasingly aware of the role he was playing as my employee for one hour. The hegemony of capitalism over the seemingly private corrupts this relationship, reducing possible friends to employees and clients, conversations to commercial transactions. Everything, even art, has its price. Rimini Protokoll has created an intercontinental phone play that isn't just about global capitalism, but that actually forces you to *participate* in global capitalism. Through the hour-long experience, I came recognize my place high up in the globalized, capitalist hierarchy, exploiting my power over Das. But simultaneously, he helped his employer to exploit me, turning my ignorance of the room and its electronics into his pay-day, which in turn allows him to employ a servant of his own.

The closer one examines this relationship, the more complicated it becomes. Rimini Protokoll provides no way to actually confirm the identity of the call center operators, or their "employer" Descon Limited. And you can forget about authenticating the individual biographies the call center employees claim as theirs. Das spoke conversational German, and told me he was recruited for the project at the local Goethe-Institut. These facts, if true, along with the picture of his servant, make it likely that he comes from an upper-middle class family. In other words, he probably was not a regular employee of the Descon call center. This is unsettling, to say least. If Das isn't "real," if he is a performer *mimicking* a call center operator, then my experience of learning and interconnectedness was, perhaps, worthless—especially if I *believe* it was worthless. This could also be frightening, a manifestation of the loss of boundaries between the public and private, between what we intend for people to know about us through the internet, and what they can actually find out. Knowing intimate details of the room and even the area beyond the window, the "call center operator" could very well be in the same building. And it casts an ethical shadow over the event—a violence done by Rimini Protokoll in hiding the voices of actual call center operators. Rimini Protokoll undercuts feelings of closeness (to the human laborer in a giant system) with these uncertainties.

So where do we stand after acknowledging the full range of contradictions embedded in both *Wallenstein* and *Call Cutta in a Box*? Perhaps as a means of approaching this question, we should turn to Bertolt Brecht. The mid-twentieth century theater director and theoretician stands behind much of Rimini Protokoll's work. Rimini Protokoll creates sly, latter-day *Lehrstücke* (learning-teaching plays). The group uses techniques drawn from documentary theater—having real, or at least questionably real, individuals tell their own stories—as well as the *Verfremdungseffekt*—estranging everyday experiences through the theatrical to allow us to see them anew. *Call Cutta in a Box* especially is a teaching or learning play in the spirit of what Brecht originally intended, with audience members taking roles and participating in the creation of a story. Rimini Protokoll uses Brechtian techniques to expose the laborer at the call center, as well as the artists' labor in creating the piece. They allow multiple meanings to assert themselves at the same time—revealing how tied up we are in the amoral economics of capitalism, while also utilizing new modes of connection made possible by globalization to bring disparate people together.

With the proliferation of media and modes of communication and entertainment, our concept of the "real" in theater and in life has changed. In the theater, at least since the advent of film, the "real" has often been synonymous with the "live"—bodies sharing space and time have been given more ontological weight than video (as can be noted in Peggy Phelan's writings, quoted above). In *Wallenstein*, Rimini Protokoll challenges this by creating a tension between the non-professional "experts" and artifice. Schiller's art gives Gassner relief in coming to terms with his failed relationship; the scenarios in Reich's officer training rupture his ability to differentiate the real and the performed, while causing lasting psychological damage. Meanwhile, we come to distrust their intentions and the truth of their stories—especially with Otto, the politician (read: professional performer). In *Call Cutta in a Box*, Rimini Protokoll pulls participants close to the call center operators using technology (voice and video Skype) and conversation in making the experience feel "live" even without bodies sharing space. They simultaneously complicate the relationship's validity (its possible artificiality) and ethicality (how the West represents and does business with India). And they do so using that same technology (the layers of Skype video; the questions about who you're really talking to) and conversation (the work's clear construction by Rimini Protokoll).

Baudrillard, like Plato and others before him, was notoriously skeptical about performance, as well as media in general—in the theater, people *try* to dissemble, to convince audiences of the truth of the mask. As Baudrillard says, "the evil spirit of simulation" leads to the even worse "evil spirit of incredulity."[14] *Wallenstein* and *Cull Cutta in a Box* are not tragedies in which the real collapses into the virtual. They do not rely on the simplistic notion of non-professional "experts" as exempt from these philosophers' problems with mimesis. Nor do they present the "experts" as unambiguously real. (Baudrillard condemns this in the form of an early version of reality television, as not only "utopian," but also "an aesthetics of the hyperreal."[15]) Instead, Rimini Protokoll relies on and promotes its audiences' ability to discern the layers of "truth" and simulacra, to critique media and individual self-presentation—in the theater, in technology-rich economy-rooms, and on the street. Rimini Protokoll embraces, and asks us to embrace, multiple modes of sight simultaneously, revealing and reveling in the paradoxes of our mediatized lives. By participating with their performances (as an active spectator or as an actor), you see and interact with art, hyperreal simulations, and the "real" using your eyes, your body, your imagination, and digital screens and cameras—all are compromised and insufficient, but also transformative and illuminating.

Notes

Portions of this essay were originally published as "Chat Room: Rimini Protokoll's *Call Cutta in a Box*" in *PAJ: A Journal of Performance and Art* 94 (2010). Special thanks to MIT Press for the permission to republish.

1. *Wallenstein. Eine dokumentarische Inszenierung*, created by Rimini Protokoll (Stefan Kaegi, Helgard Haug, and Daniel Wetzel), 2008, DVD.
2. *Call Cutta in a Box*, created by Rimini Protokoll (Stefan Kaegi, Helgard Haug, and Daniel Wetzel), Goethe Institut, New York City, NY, February 9, 2009.
3. Frederic Jameson, *Postmodernism, or, the Cultural Logic of Late Capitalism*, (Durham, North Carolina: Duke University Press, 1991), xi.
4. Walter Benjamin, "Theses on the Philosophy of History," *Illuminations*, ed. Hannah Arendt, trans. Harry Zorn (New York: Schocken Books, 1969), 258.
5. In 2006, Rimini Protokoll turned this interruption into a stand alone, twenty minute long production, *Männer meldet euch!* [Calling all Men!, Kunsthalle Vienna], starring Mischereit alone on the stage taking calls from prospective costumers.
6. Jean Baudrillard, *Simulacra and Simulation*, trans. Sheila Faria Glaser (Ann Arbor: The University of Michigan Press, 1994), 10.
7. Ibid., 11.
8. Ibid.
9. Mischereit: "Sie haben Gluck gehabt mit diesem Vers, denn *Wallenstein* ist eigentlich langweilig." Gassner: "*Wallenstein* ist trostlos."
10. William Shakespeare, *The Winter's Tale*, ed. Frances E. Dolan (New York: The Penguin Group, 1999), IV.4.95-96.
11. Peggy Phelan, *Unmarked: The Politics of Performance* (New York: Routledge, 1993), 41.
12. Philip Auslander, *Liveness: Performance in a Mediatized Culture* (New York: Routledge, 1999), 8.
13. Gayatri Spivak, "Can the Subaltern Speak?," in *Marxism and the Interpretation of Culture*, ed. Cary Nelson and Lawrence Grossberg (Urbana: University of Illinois Press, 1988), 271-313.
14. Baudrillard, *Simulacra and Simulation*, 107.
15. Ibid., 28.

Chapter 16

Nazis and Nazi Costumes: The State of Evil in Hyperreal Ethics

Brian Johnson

Jean Baudrillard's ideas about the hyperreal, as expressed in *Simulacra and Simulation* (1985), suggest a state of culture in peril. Part of the reason that his work takes this tone is that, under the heading of the hyperreal, he describes a system that breaks away from standard assumptions about constructing new cultural values from old by affirming or opposing prior positions, and suggests that postwar culture has developed a new system for evaluating value itself. His is not a critique of the methods by which individual values are sorted into conditions like shocking or banal, right or wrong, dangerous or safe, but a critique of the method of sorting. By providing a survey of disparate cultural artifacts, including religious laws concerning graven images, Disneyland, and Watergate, he demonstrates that the criteria of evaluation and the act of evaluation have become inseparable.

Were Baudrillard's hyperreal only a commentary on evaluation in a general sense, it might suggest an avant-garde method for the production of cultural value, perhaps frightening but, undoubtedly, also filled with potential. The problem that inspires the dystopian motif for the hyperreal is that the restructuring of evaluation is total. Baudrillard's point is that it is not the particulars of the culture that are dangerous; his wide reading is a means of commenting on the method by which a culture constructs its elements—including, ultimately, the values belonging to ethics. The end result is a culture that cannot tell the difference between right and wrong.

What the hyperreal suggests for ethics is that there can be no stable, pre-experiential ethical values against which we might compare our own experiences for authenticity and certainty. Instead of Laws set into the buckram of existence or handed down from a divine figure, the culture creates laws which are arbitrary and contextual. Following these laws is not necessarily good, but lawful or appropriate, and violating them is not so much evil as it is a transgression. In this hyperreal dystopia, good doesn't struggle against evil for the fate of the world; instead, good and evil fade into myth. "The fetishized history will preferably be the one immediately preceding our 'irreferential' era. Whence the omnipresence of fascism and of war in retro—a coincidence, an affinity that is not at all political."[1] Paradoxically, even as the fetishizing reduces the political meaning of fascism, the drive towards morality (and against immorality) engendered by the fetish ceases to be anything but a political construction. Good and evil achieve definition through a navigation of value by the *polis*.

Baudrillard posits as the impetus for this change a kind of nostalgic hatred of fascism because it is this very nostalgia that drives his description of the hyperreal moral system. In doing so, he places fascism and its most public symbol, Nazism, in a unique position. An analysis of the effect of Nazism within the realm of hyperreal culture requires a wide reading of the body politic similar to the one supplied by Baudrillard. Thus, the ubiquity of the Nazi as a hyperreal construct is demonstrable only by a cultural survey—an analytical mode which tends to work against the disciplinary framework of close reading in a manner reminiscent of Baudrillard's wide reading in his description of the hyperreal.

In pointing out the importance of fascism to the fundamental philosophical shift of the mid-twentieth century, he is, of course, not alone. Lyotard also points to the great crimes of the Nazis as the inaugurating event for the skepticism of Postmodernism. He writes:

> [T]he silence that the crimes of Auschwitz impose upon the historian is a sign for the common person. Signs are not referents to which are attached significations validatable under the cognitive regimen; they indicate that something which should be able to be put into phrases cannot be phrased in the accepted idioms.[2]

Thus, for Lyotard, Auschwitz and the attempt, not only to murder its prisoners, but also to erase their history becomes the impetus for a new discursive system, one which has idioms to describe absence of meaning. His primary metaphor is that of an earthquake that destroys the instruments that might measure it. The sins of Nazi Germany are so heinous that, after their revelation, there can be no evil able to measure up to them for comparison. Thus, all lesser evils are muted out in their moral significance and are made subject to context dependent evaluation or else comparable only to Nazi Germany for a stable definition of evil.The latter manner of measuring evil is critiqued by numerous other sources as well: including most notably George Orwell and Leo Strauss. Just after the end of World War II, Orwell wrote "the word Fascism has now no meaning except in so as it signifies 'something not desirable.'"[3] Orwell's sentiment is echosed by Strauss's "*Reductio Ad Hitlerum*" coined in 1953 by Leo Strauss in *Natural Right and History*:

> Unfortunately, it does not go without saying that in our examination we must avoid the fallacy that in the last decades has frequently been used as a substitute for the *reductio ad absurdum*: the *reductio ad Hitlerum*. A view is not refuted by the fact that it happens to have been shared by Hitler.[4]

Together, Orwell's complaint and Strauss's coinage suggests that the problems associated with comparisons to Nazism have plagued American ethics since the beginning of World War II. Strauss's fallacy is repeated, tongue-in-cheek for the information age, as "Godwin's Law," coined by Mike Godwin and in popular usage, which suggests that any argument on the internet, if allowed to run long enough, will eventually invite a reference to Nazism or Hitler.[5]

Baudrillard differs significantly from Lyotard, Orwell, and Strauss precisely because he recognizes that the draw of fascism is not simply an occasion for the deconstruction of differentiating criteria. The reign of the Nazis in Germany represents a particular time when evil was clearly defined. For Baudrillard, the Third Reich becomes a point in history (or its hyperreal counterpart) that anchors the arbitrary evaluations of hyperreal ethics and gives them a basis of authority. To be like a Nazi then becomes synonymous with being evil, but the exclusivity of the Nazis in this role is wholly dependent upon their being the only evil. He writes: "Fascism's cruelty, its terror is on the level of *this other terror that is the confusion of the real and the rational*, which deepened in the West, and it is a response to that."[6] In response to Nazism, only a similarity to the Nazis is allowed the same ethical value as being evil—and it is only because of the evil of the Nazis and their reliance on stable values that an opportunity for validation and endorsement of an alternative method for value making arises (one in which the real and the rational are confused). According to Baudrillard:

> Forgetting extermination is part of extermination, because it is also the extermination of memory, of history, of the social, etc. This forgetting is as essential as the event, in any case unlocatable by us, inaccessessbile to us in its truth. This forgetting is still too dangerous, it must be effaced by an artificial memory.[7]

This artificial memory is the hyperreal version of the Holocaust and Nazism. Using Lyotard's metaphor, the earthquake not only destroys the instruments that measure it but, somehow, also destroys the event of the earthquake. In its place is only a made for TV rendition of the disaster.

Baudrillard recognizes that this definition of Western postwar culture as a reaction to Nazism is, unfortunately, as deeply affected as any other hyperreal simulacra by the tension between what might be called historical truth and its representation in popular depiction within American mass culture. Nazism, under the rubric of the hyperreal, is just as prone to the confusion of message and medium. The reason for this dynamic is that, though the various military actions and infamous domestic policies of Nazi Germany are well documented, the actual meaning of Nazism is deeply affected by the artifices of culture like film, television, literature, political propaganda, and official state policy.

What Americans understood about the Nazis even on the eve of World War II, for instance, was reflected by commercial films like *The Great Dictator* (1940) and *Casablanca* (1942), which were soon followed by American propaganda like the *Why We Fight* (1943) series directed by Frank Capra. At the same time, the relationship between the real Nazis and their depiction were not (and is not) one directional—*Casablanca* may have reflected existent cultural values, but it also informed its contemporary viewers as to why the Nazis were evil and why that evil demanded Americans abandon their isolationism. Rick Blaine offers his philosophy early on in the movie when he says, "I stick my neck out for no one," to which his competitor Ferrari offers, "when will you realize in this world today, isolationism is no longer a practical policy."[8] The film acts as a chronicle of Rick's transformation into the kind of American who becomes involved in the fight against Nazism because he cannot, in good conscience, stay on the sidelines. *Why We Fight*, as a state-sponsored work of propaganda, was meant to not only reflect existing values but to affect them as well: to create an image of the Nazi enemy so imperialistic that war was required.

Since the end of the Second World War, the defeat of Nazism, and the revelation of the Nazi atrocities, other issues related to the Nazis as evil have likewise been affected by their popular representations. The public's understanding of the Holocaust, for instance, has been affected by the success of *Anne Frank: Diary of a Young Girl* (1947), the 1978 *Holocaust* television mini-series, and later, *Schindler's List* (1993). Holocaust historians point to the global dissemination of news about the trial of Adolf Eichmann as a seminal moment in modern Holocaust consciousness nearly twenty years after the end of World War II. The American public's first understanding of the Holocaust itself came from newsreel footage of camp liberation in 1945—films which notoriously underplayed the ethnicity of the victims and presented camp prisoners in terms of political and national affiliation.

Likewise, the history of World War II as it is understood by the culture at large is deeply affected by World War II combat films. The highest praise one might offer *Saving Private Ryan*, for instance, is that its depiction of Omaha Beach was unapologetically realistic: the film finally serves as an adequate basis for constructing assumptions about the historical event. More recently, Quentin Tarantino's film *Inglourious Basterds* (2009) completely rewrites the end of World War II to create a more vengeful version of history in which the Nazis receive proper punishments for their crimes. Thus, simulation and simulated mix, and the lessons of history become the lessons of history as depicted in film.

Jean Baudrillard anticipates this confusion of history and historical depiction: "the territory no longer precedes the map, nor does it survive it. It is nevertheless the map that precedes the territory—precession of simulacra—that engenders the territory."[9] In terms of the hyperreal map of Nazism created through depictions of that ideology (and its history) in popular culture, what Americans believe to have actually happened becomes an amalgamation of canonical history, film and television shows all related to what historically happened and each affecting each other as well. What American culture presents to itself as a prime example of evil is preceded by the simulacrum of Nazi evils both on screen and suffused throughout the culture—a precession that includes history, historical fiction, and pure fiction. In the 1977 Bakshi animated film *Wizards*, for instance, the evil wizard Blackwolf uses old Nazi propaganda films to call evil faeries to war in a post-apocalyptic world 10,000 years in the future. He leads his forces into battle screaming, "It's time to strike! Sieg heil!"[10] Where and how does the culture rearrange its understanding of Nazi Germany so as to make room for Blackwolf?

The question is troubling precisely because Blackwolf is not alone. Because of its fascination with the issues surrounding Nazi Germany, American culture produces hundreds of references to Nazism a year in an amazingly disparate set of contexts, but because the subject is accepted as having supreme levity, few critical frameworks are available to help discern their implications. The disciplinary frameworks collected under the heading of Holocaust Studies may, for instance, provide ample material for understanding *Schindler's List*, but include little critical framework to help analyze the fiction of Nazism: for instance, the anti-Semitism of Eric Cartman in *South Park*, the clockwork Nazi assassin Karl Ruprecht Kroenen in the *Hellboy* movie (2004), or *The X-Men* villain Magneto as a Holocaust survivor.

Likewise, there are scant few analytical apparatuses to analyze the common political comparisons to Nazi Germany like the numerous such indictments made against Presidents Ronald Reagan, George W. Bush, or Barack Obama. What does it mean, ultimately, when Americans come to recognize Iran as a member of the Axis of Evil? What does it mean for them, what does it mean for the definition of Nazism, and what does it mean for the American culture's definition of evil? Perhaps the most famous of all American civil rights advocates, Martin Luther King, Jr. wrote concerning his resistance to state laws in his "Letter from a Birmingham Jail," "We should never forget that everything that Adolf Hitler did in Germany was 'legal' and everything the Hungarian freedom fighters did in Hungary was 'illegal.' It was 'illegal' to aid and comfort a Jew in Hitler's Germany."[11] Must the laws of segregationist South be labeled Nazi-like in order for their condemnation? Isn't the terror of lynching enough to warrant outrage? In Alexander Cockburn's condemnation of Abu Ghraib in his column in *The Nation*, he makes a direct reference to Nazi Germany as the precursor to the torture tactics used by the U.S. Government. The reference occurs as a descriptor in his statement: "the long and copiously documented record of U.S. torture, with many of its refinements acquired by the CIA from the Nazis after World War II."[12] He goes on to describe prisoners being raped with foreign objects, kept awake all night in cells, led around on dog leashes, having their genitals threatened by dogs, and finally, executed. Is it really necessary to know that this is reminiscent of Nazi Germany in order for the public to appreciate that such torture is wrong? In all these cases the overriding question becomes whether or not these acts have so little ethical value in their own context that they should need the crutch of a Nazi metaphor.

Yet, in a hyperreal amalgamation of culture, all of these sources engender the definition of Nazism as much as, or more than, canonical history, precisely because they require its stability. Taken together, these fictional, political, and rhetorical depictions of Nazi Germany produce meaning for Nazism supplemental to references which seem to carry more authenticity. *The Pianist* may, perhaps, have more authority than *Office Space* on the horror of the Holocaust, but unrealistic sources, if popular, have their effect too. The definition of historical Nazism isn't entirely free from the effects of *Indiana Jones.*

These opportunities for analysis afforded by these popular sources cannot reside within the providence of Holocaust studies, because they have little to say about the real genocide or the real Nazi Germany. If we view these moments as commenting on some pretend place, a pastiche or a parody of the historic, then we are forced to find some new mode of analysis that is parallel to Holocaust studies, but separate from it: a hyperreal Holocaust study or a hyperreal philosophy of ethics to match a hyperreal history—one which answers questions about how a culture utilizes depictions of catastrophe and inhumanity to generate ethical value. Still, the separation between the real Nazis and their hyperreal counterparts is difficult to endorse as the distinction between the real as a model and the fabricated as derived from the model are not so easily delineated. As Baudrillard posits:

> The medium is itself is no longer identifiable as such and the confusion of the medium and the message is the first great formula of this new era. There is no longer a medium in the literal sense: it is now intangible, diffused, and diffracted in the real, and one can no longer even say that the medium is altered by it.[13]

Thus a concentration on fabrication is not completely freed from the concerns or methods of a study of an historical truth (like Holocaust studies), but instead, seems to work at similar problems from an entirely unorthodox position which challenge, as Baudrillard suggests, "the difference upon which the law is based."[14]

About such a mythology, Baudrillard argues, "We require a visible past, a visible continuum, a visible myth of origin, which reassures us about our end," but one might just as well say that the mythology built up around the Holocaust reassures the layperson of their means: this is how evil may be recognized.[15] As long as the culture thinks it knows what happened during the Holocaust, it thinks, also, that it knows the face of evil, and is given agency about how and when to condemn it. The culture knows to dismiss the Nazis as evil, but cannot so easily differentiate between the real Nazis of history and the actors whose Nazi costume uniforms and thick German accents now inform us about the ideological and ethical meanings of Nazism, the drive to world conquest, racism, fascism, and genocide. *Schindler's List*, for instance, generally considered the touchstone film about life in the death camps, is a film based on a book (*Schindler's Ark*) which won the Booker Prize in Fiction in 1982. The novel can now safely be called a work of creative non-fiction (a genre that wasn't recognized at the time *Schindler's Ark* won the award), but this still begs questions about authenticity and authority: how much of the work is creative, how much of *Schindler's List* is a creation of its director Stephen Spielberg—and more importantly, can the audience tell the difference? Does the audience recognize that the Plaszow camp in the film is a forced labor camp, and not a death camp—that the conditions there do not, necessarily, translate over to places like Treblinka or Auschwitz where prisoners were systematically murdered? Or is one camp depiction enough to act as the simulacra for all future simulations?

The condition of hyperreality suggested by Baudrillard proves useful for wading through these effects precisely because it offers a basis for establishing and resolving concerns about governing the truth and implications of these depictions. For instance, under a methodology devoted to describing the hyperreal, the truth and validity of the depictions (a contentious point for Holocaust and cultural studies) is of little matter. They are "all true, in the sense that the truth is to be exchanged, in the image of the models from which they derive, in a generalized cycle."[16] In other words, these depictions are true in that there is no "stable truth" to which an appeal can be made for verification, no epitomized model that acts as a proof for comparison, and thus, no way to prove them false. They are, instead, compared to other simulations in an immense system composed exclusively of fictions that act upon each other to create a value which will pass for truth. Their validity is only that they may be included in such a system, but the hyperreal threatens the very possibility of any higher truth beyond the examples.

Ultimately, this system of simulation takes on a dystopian tone precisely because, as Baudrillard posits, it "is infinitely more dangerous" than a system of ethics which invites either submission or transgression, "because it always leaves open to supposition that, above and beyond its object, law and order themselves might be nothing but simulation."[17] Thus, the threat is derived from the supposition that Nazis *are* evil, but because they are also simulations, the apparent stability of this definition of evil, as derived from Nazism, may also be just another simulation.

Furthermore, as Nazis are as prone to simulation, they are as prone to being reordered according to hyperreality's recursive system of redefinition: the definition of Nazism is preceded by simulacra. If Western culture is a reaction against Nazism, it becomes increasingly a reaction against its own phantasmal notions of Nazism and not the real thing. Real Nazism (historical Nazism) becomes increasingly inaccessible, and values, without anchor, tend towards the arbitrary.

Similarly, American culture has come to accept a vision of Nazism unfettered by the demand for historical accuracy. If the Nazis need to be Satanists, then they will be depicted as such. If they need to be homosexual, then they will become homosexual. If they need to be moronic, addicted to drugs or sex, authoritarian, totalitarian, or even communists, they will take on whatever costume they need to wear. They will mutate, not according to the misrepresentation of storytelling, but because the image no longer relies on factual accounts. The stories have produced models of Nazi-like behavior, the models have produced Nazis, and these hyperreal Nazis proceed to become a new model for further portrayal of Nazis. When American protesters wave enlarged pictures of the body piles of Bergen-Belsen so as to equate national health care with the Holocaust, they are not deranged—they are right on queue.

It's easy to posit, for instance, that at the beginning of this Baudrillardian "precession" stands a figure like Major Strasser from *Casablanca*: a Nazi who appears to be the Nazi we think of as historical but Strasser is problematic. He bears no relationship to Jews whatsoever. His only concern is for world domination; Strasser doesn't want genocide, he just wants to goose step through New York. If the image of Nazism was built on Strasser initially then it began as a misinformed simulation. But accuracy has not improved since, despite public access to Nazi evils. More contemporary representations of Nazis include the Soup Nazi of *Seinfeld* fame who is like a Nazi only because he is authoritarian, and the victims of 9/11 who, according to Ward Churchill, were "little Eichmanns" who "formed a technocratic corps at the very heart of America's global financial empire"—American business as Nazism.[18] In the summer of 2009, Aldo Raine and his band of Inglourious Basterds beat in the heads of German soldiers—providing punishment for a holocaust that wouldn't be revealed, historically, until after the war, and against those for whom, historically, would bear no responsibility. In Quentin Tarantino's version of Nazi Germany, every German knows, every German is responsible, and every American action is vindicated as an act of vengeance for the victims of the death camps. Clearly Nazis no longer need iron crosses, SS runic insignias, gas chambers, or crematoriums in order to be evil. They needn't be blind bureaucrats or racists. They most certainly do not need to be German. In short, they needn't be like the historical Nazis at all.

The crux in this argument is that, as the Nazis become chimerical to the point of unrecognizable, they continue to maintain their Nazism in some fundamental way in that they remain definitive of evil. It's impossible to ask what the Nazi costume looks like at this point because it is now composed of so many various features that its list of descriptors is infinite and its possibility invades all rhetorical occasions at once. There is no milieu or context in which one can be shocked to find a Nazi reference lurking.

The deterioration of the Nazi's stable ethical position might mean very little if that position weren't unique, but it is only the Nazi that serves this particular ethical function (impetus to the hyperreal and symbol for the tragedy of its excesses). Nazis are employed as an evil that can operate outside of context because of its understood meaning inside of American mass culture: nothing is worse than the Holocaust and no contextualizing of the genocide will make it any less horrible.

If, however, Nazism is the lynch pin of hyperreal ethics, and the costume of Nazism fails to refer to an evil outside of context, then the dystopian tone of the hyperreal is warranted. Baudrillard describes the culture's capacity to differentiate between right and wrong as an aesthetic, as opposed to moral, value. The context-independent definition for evil, upon which the entire stability of the irreferential hyperreal ethics depends, is a myth: Nazis are as prone to context dependency as anything else, and thus, the definition of evil becomes meaningless in this most extreme case, and in all lesser cases that turn to Nazism in comparison. The fixity of Nazism as a definition causes all other meanings to become unfixed, and then it too unfixes itself, and the undifferentiated system, turned to in reaction to fascism, loses all value whatsoever.

Notes

1. Baudrillard, Jean, *Simulacra and Simulation* (Ann Arbor: The University of Michigan Press, 1994), 44.
2. Lyotard, Jean-Francois, *The Differand: Phrases in Dispute* (Minneapolis: The University of Minnesota Press, 1988), 56.
3. Orwell, George, *The Orwell Reader: Fictions, Essays, and Reportage by George Orwell*, (New York: Harcourt, Brace and Company 1956), 359.
4. Strauss, Leo, *Natural Right and History* (Chicago, University of Chicago Press, 1953), 42.
5. Skirvin, Tim, "How To Post About Nazis and Get Away With It—The Godwin's Law FAQ," *Internet FAQ Archive*, 2/15/09, http://www.faqs.org/faqs/usenet/legends/-godwin/.
6. Baudrillard, *Simulacra and Simulation,* 48.
7. Ibid. 49.
8. *Casablanca*, DVD, directed by Michael Curtiz (1943; Warner Bros., 1999).
9. Baudrillard, *Simulacra and Simulation,* 1.
10. *Wizards*, DVD, directed by Ralph Bakshi (1977; 20th Century Fox Home Entertainment, 2004).
11. Young, Richard P. *Roots of Rebellion: The Evolution of Black Politics and Protest Since World War II.* (New York: Harper and Row, Publishers, 1970), 337

12. Cockburn, Alexander, "Green Light for Torture," *The Nation*. May 31 2004.
13. Baudrillard, *Simulacra and Simulation,* 30.
14. Ibid. 21.
15. Ibid. 10.
16. Ibid. 17.
17. Ibid. 20.
18. Churchill, Ward. *On the Justice of Roosting Chickens* (Oakland: AK Press 2003).

Chapter 17

Operational (Hyper)realities in the Exilic Labyrinth: Roberto G. Fernández's Construction and Destruction of Identity through Parodic Simulacra

David de Posada

The first wave of Cuban refugees who came to the United States, the exodus between 1960 and 1973, allowed for many of these immigrants to reinvent themselves in a new country and cultural environment.[1] Often triggered by nostalgia, a desire to reestablish a lost comfortable social position and a sense of cultural belonging or connection, the Cuban exile attempts to anchor the present through a romanticized past where recreation and simulation are recurrent themes. Jean Baudrillard's postmodern theories of simulacra, where the human experience is of a simulation of reality rather than reality itself, allows us to establish parallels with the Cuban exiles' excursions into what ultimately falls under Baudrillard's third order of simulacra, where a distinction between reality and representation is often indistinct.

Roberto G. Fernández often places his characters—using the first wave Cuban exile as inspiration—in an antagonistic space full of social and cultural challenges where they are constantly reminded of their displacement. Unlike most other U.S. ethnic literatures where characters' victimization by the hegemonic culture is so prominent and an essential element in character development, Fernández's personages are not portrayed as defenseless victims of an oppressive society.[2] Coping mechanisms, notably nostalgic escapism, allow many of his characters to function in what could be described as hyperreal environments: realities that are not simply imagined but rather simulated, recreated and ultimately fused with the real. For Baudrillard, this simulation results in the "hyperreal" and is, "[t]herefore, pretending, or dissimulating, leav[ing] the principle of reality intact: the difference is always clear, it is simply masked, whereas simulation threatens the difference between the 'true' and the 'false', the 'real' and the 'imaginary.'" In *Simulacra and Simulation*, Baudrillard quotes Littré's simile of the illness in which "whoever fakes an illness can simply stay in bed and make everyone believe he is ill. Whoever simulates an illness produces in himself some of the symptoms."[3] Baudrillard uses Littré's example to take the idea of simulation to the next level, the subconscious:

> As to psychoanalysis, it transfers the symptom of the organic order to the unconscious order: the latter is new and taken for "real" more real than the other—but why would simulation be at the gates of the unconscious? Why couldn't the "work" of the unconscious be "produced" in the same way as any old symptom of classical medicine? Dreams already are.[4]

Fernández's best developed characters, and thus those most often discussed by critics, embody this idea of a psychosomatic reality through a fabrication of an adopted identity. Nellie and Mrs. Olson in *Holy Radishes!* or Mirta's recreation of Varadero Beach in her bathroom and her hyperbolic Paradisial fantasies about Cuba in *Raining Backwards*, for example, cannot simply be dismissed as parodic caricatures.[5] It is essential to take into account that while Fernández portrays the Cuban exile through parodic hyperbole, the characters' creations, recreations and nostalgic recompositions also produce in them symptomatic manifestations in the form of memories. These memories, as the characters want to remember them, are created in a world where fantasy and reality are fused and often indistinguishable. The impossibility of confirmation or referent due to a temporal and geographical disconnect maintain these fantasies in a realm of fabricated reality; again, the hyperreal. Nostalgia becomes then, as Baudrillard underlines, an enhanced reality:

> When the real is no longer what it is, nostalgia assumes its full meaning. There is a plethora of myths of origin and of signs of reality—a plethora of truth, of secondary objectivity, and authenticity. Escalation of the true, of lived experience, resurrection of the figurative where the object and substance have disappeared.[6]

The need for recreation of an idealized past and of an identity as a necessary coping mechanism allow temporal and physical spaces to superimpose—thus masking the harsh reality of an uncooperative foreign and incongruent environment and making it tolerable. The act of remembering and retelling creates each time a new version of a reality that is farther removed from the real. This new version becomes the hyperreal: "the generation by models of a real without origin or reality."[7]

As I argue in "Operatic Transposition and the Romantic Aesthetic in the Works of Roberto G. Fernández," the Cuban exiles' escapist reality echoes that of the dejected heroine in the operatic mad scenes of the nineteenth century. In the Romantic tradition, the forlorn character escapes the harsh reality of her incongruent and hostile environment by recreating and returning to a much happier state.[8] In the context of the Cuban community, this escapist coping mechanism could be seen as a metaphorical mental illness; a collective mad scene of sorts that drives this group from an oppressive environment—an exilic world perceived by the immigrant as laden with cultural and social incongruence—into a new operational reality. I establish this parallel to underline that the creation of reactionary spaces and that of operational altered states could offer a different angle in understanding the Cuban exile's restorative environments where the real often becomes blurred by a distorted and often hyperbolic recreation. This discussion, however, focuses on how other symptomatic elements support this argument of an altered reality and bring the subconscious to be mirrored and sustained by tangible, recreated realities. Baudrillard's ideas on the real support this concept that what may seem irrational could also be perceived as a distilled version of the real—capturing an essence that facilitates new operational systems and thus gaining its own reality. As Baudrillard explains, "[The real] no longer needs to be rational, because it no longer measures itself against either an ideal or negative instance. It is no longer anything but operational. In fact, it is no longer really the real, because no imaginary envelops it anymore. It is a hyperreal."[9]

This idea of the hyperreal is best captured in Fernández's short story, "Kon-Tiki," where the author synthesizes the power and danger of reinvention through what could be described as parodic simulacra.[10] In this hyperbolic state of imitation, a copy is achieved to perfection, albeit with disastrous consequences that underline the author's penchant for highlighting the ironic and futile attempts at cultural integration so common among the Cuban exilic community. This precarious link, described by Pérez Firmat in *Life on the Hyphen*, remains that place of fragile connectivity that actively threatens to separate Cuban-Americans from their adoptive culture. In "Kon-Tiki," a group of illegal immigrants from various Latin American countries, all living in fear of being deported to their respective countries, join forces to craft a plan to fool the American authorities and remain in the United States through the "wet-foot dry-foot policy."[11] Their very intricate ruse requires that a Honduran, a Haitian, a Mexican and a Colombian adopt Cuban identities, to reinvent themselves and become perfect copies of Cuban refugees through imitation.

Their plan calls for the adoption of new names, accents and personal histories through careful memorization and coaching, all to be accomplished by following the strict guidelines of a manual. The narrator, a Mexican, completely foreign to Cuban culture, explains the initial difficulties in assembling the necessary elements to produce the copies:

> The only setback I had at the very start was the compilation of the teaching manual. The manual included vital information on the country for us to learn. I was gathering data from the old lady that lived in the pink house with the green verandah. Luckily, I realized that she lived on fantasy island, a place that had ceased to exist forty five years ago, and furnished the accurate information for the future crew.[12]

To the reader familiar with Fernández's works, the "old lady" supplying the information is undoubtedly Mirta, from *Raining Backwards*, an allegorical character that reappears in several of Fernández's works—a leitmotif that reminds the reader of the distorted reality created by nostalgia, recreative fantasies and re-membering. This reference foreshadows disastrous consequences and prefaces a recreation that is essentially doubly faulty.[13] Because the narrator identifies the supplier of information as unreliable—based on his perception that the "old lady's" account is tainted by excessive manipulation—he recreates his own version of reality taking the supplied narrative into the realm of the hyperreal. While Mirta does not explicitly appear in "Kon-Tiki," it is important to underline that Fernández builds this character, in *Raining Backwards*, around a hyperbolic notion of nostalgia allowing her to weave a cultural and social mythology that is progressively distanced from reality. Jalena Sesnic describes the flow of memories from the middle-age Mirta to her teenage neighbor, Eloy, in *Raining Backwards* as "an exchange . . . [that] works not only on the level of transmission through their narrativization. Mirta remembers, but also through her stories recreates something that never was."[14,15] In "Kon-Tiki," the narrator's reinterpretation, devoid of any authentic cultural mapping (again, he is Mexican and has never been to Cuba), distances reality even further from any remnant of truth. The Mexican narrator's manipulation of the Cuban referent in "Kon-Tiki," like Mirta's transmission of a mythological nostalgia to Eloy in *Raining Backwards*, becomes the vehicle by which (hyperreal) identities are created and infused to the group of illegal immigrants.

As the events in "Kon-Tiki" unfold, it becomes clear that while authenticity of an original is a desirable requirement for the creation of a perfect reproduction, it is not vital however, in the creation of the hyperreal. Once the participants are reprogrammed and adopt their Cuban personas, the process of recreation and simulacra requires the manufacturing of objects or tangible realities to support the idea of authenticity and the creation of what Umberto Eco calls "connective tissue": the seamless integration of the fake and the real to create hyperreal environments. The careful attention to detail in the fabrication of the raft, for instance, creates the desired illusion of a desperate resourceful creation as explained by the narrator:

> The plan had worked out without a hitch, and the equipment of the vessel had taken thirteen days. Granted, it could have been done sooner, but I'm a perfectionist. The goal, after all, was authenticity . . . The raft looked perfect, with a real third world quality to it.

As Eco points out in his description of a wax museum's representation of the Oval Office, "for historical information to be absorbed, it has to assume the aspect of a reincarnation. To speak of things that one wants to connote as real, these things must seem real. The 'completely real' becomes identified with the 'completely fake'. Absolute unreality is offered as real presence."[16] For these illegal immigrants, authentication of the fake becomes the purpose, the fundamental goal, initially presenting an accomplished new reality. Unfortunately for this group, things go terribly awry when their raft capsizes and their adoptive identities betray them. When intercepted by the coast guard, the illusion was so complete, so real, that as "wet feet" they are sent back to Cuba.

The fabrication of identities, altered realities and simulacra become the more necessary in an environment of displacement. In *Holy Radishes!* Fernández's second novel in English, the author's use of parody serves to emphasize the tragic consequences of social incongruence by reminding the reader of the futility of attempting cultural assimilation. Two of the main characters, Nellie, a Cuban who has fabricated an Italian persona and believes to have been born in Italy and Mrs. James B., her American neighbor, coworker and friend—both living in a small South Florida town—find themselves in an awkward place between peaceful coexistence and the constant discordant notes produced by their cultural and social disparities. Nellie, who finds herself continuously and desperately seeking acceptance and a sense of grounded identity, asks: "But are you still my friend?" to which Mrs. James B. replies with pragmatic superficiality: "I'll be a better friend if we get rich."[17] Both trapped in suffocating environments, with disillusionment as a common factor, the friends embark in a joint venture to make money in anticipation of imminent unemployment. Their plan calls for inventiveness and the creation of a hyperreal environment:

> The packing house might close down, which means we might be out of a job by Christmas, which means we have to think fast, which means I already thought fast. . . [l]isten to what's gonna make us richer than Rockefeller . . . we're going to build us a South American mini-zoo.[18]

After capturing a series of common animals, and collecting scraps from the dumpsters of a local seafood market, butcher shop and hardware store, the women begin their creation of the exotic creatures:

> Nellie, let's start with the bullfrog. Hand me that purple spray. Let's paint his legs. Then get me half a lobster shell and fit him inside and then strap it with the rubber band [. . .][19]

Following the manufacturing of the many hybrid animals and giving each one not only a Latin genus and species but also a geographical origin to complete their pedigree, the zoo opens and is well attended by an enthusiastic crowd. As in "Kon-Tiki," however, the manufacturing of false identities leads to disaster. The carnivalesque freak show of exotic animals sees the entire town gather to admire and marvel at their bizarre and unique appearance. The fabricated exotic display and the resulting deception come to an abrupt end after an unexpected torrential rain falls on the creatures as a disappointed and irate crowd watches them morph back into ordinary animals. If we consider Fernández's penchant for underlining social and cultural displacement, the exotic zoo could easily be perceived as an allegorical representation of reinvention, simulacra and ultimately, the failed attempt to deceive the hegemonic culture. The author also seems to take this opportunity to end this episode with a strong social critique against consumerism, American superficial values and excesses with a parody of the Great Flood:[20]

> Mrs. James B. was running from cage to cage in a futile attempt to save the creatures from further deterioration. The waters were rising by the minute, but by the time the big entrance sign started floating with the current, she had gathered most of her metamorphosed species and placed them on top of a picnic table. The board began drifting with the deluge, the flood that ultimately drowned her American dream.[21]

In contrast, and again underlining the social and cultural incongruence between the two cultures, Fernández puts Nellie in an oneiric space, seemingly so common as a coping mechanism within the Cuban exile community—the return to a better time and place.[22] The distressed Nellie responds to the angry crowd demanding their money back by drifting away to her beloved Mondovi, her imaginary place of origin and the source of her escapist fantasies:

> In her mind she saw herself dressed as a nurse distributing loaves of bread to the hungry masses from an open cart. The allies had liberated Mondovi.[23]

This escapist nostalgia enables the creation of hyperreal environments and facilitates reinvention. Vijay Agnew in *Diaspora, Memory and Identity* explains the power of remembering as an active, transformative process and quotes Gayle Green who distinguishes it from *nostalgia*, the more stagnant activity:

> Whereas "nostalgia" is the desire to return home, "to remember" is "to bring to mind" or "think of again," "to be mindful of," "to recollect." Both "remembering" and "recollecting," suggest a connecting, assembling, a bringing together of things in relation to one another. . . . Memory may look back in order to move forward and transform disabling fictions to enabling fictions, altering our relation to the present and future.[24]

This metamorphosing of fictions allows for the creation of new and hyperreal realities. Nostalgia remains the key to remembering and to this active transformation. Because nostalgia, as Green defines it, is "a form of escape from an unsatisfactory present into an idealized past," it only serves to set the mechanism of reinvention in motion and in turn has the capability to create not only new identities but also new realities.

New realities are explored by both Jean Baudrillard and Umberto Eco, who seem fascinated by holographic technology and use it as a metaphor in their discussions of hyperreality.[25] Fernández's chapter from *En la Ocho y la Doce*, "Hologramas" (Holograms), ties the idea of this virtual reality to the Cuban exile community and their longing for nostalgic recreation.[26] In this chapter, Fernández creates a futuristic product in a home for the elderly where nostalgia becomes a commodity and holographic and projected images recreate a past based on representations of faint memories. The residents' relatives are asked to furnish photographs of living and deceased family members and friends to create life-size holographic images that facilitate an enhanced reality; a hyperreal environment. The facility's manager explains how the product works:

> For the Victoria Plan we need you to send us pictures of your grandparents, great grandparents, cousins, aunts and uncles, close friends and of your father, if he has passed away . . . and from all of these we make some holograms that we project in your mom's room so she feels as though they are keeping her company . . . the next step is to select from our list of towns and cities.[27]

The elderly are immersed in familiar surroundings, where the past overlaps with the present, where temporal and geographical lines never intercept but succeed in creating a new anesthetizing reality through nostalgic overload.

While the holographic environments of the nursing home may seem a bit farfetched, Fernández reminds his readers of the power of nostalgia as captured in photographs and how these could serve as a passport for reinvention and the creation of the hyperreal. As I established earlier, Mrs. James B., the character from *Holy Radishes!,* is disillusioned. As an aging cheerleader and prom queen with an abusive husband and a world that threatens to smother her, she senses a loss of identity. After months of hearing Nellie's fabrications about her past in her distant paradise—complete with descriptions of exotic locales and fueled by nostalgic abandon and a past worthy of revisiting—Mrs. James B. decides to acquire a new past and a fresh identity by purchasing old photographs in a pawn shop.

> She sorted through the pile, took a few and discarded the rest, creating her past at will. . . . Mrs. James B. continued digging in the pile, at times feeling like an archeologist. When she was finished, she had bought enough pictures to have a three-generational family . . . including a couple of very recent snapshots of

two children . . . five pictures of the mansion, eight portraits of a lady and five of a gentleman.[28]

It is in this chapter, "The Pawn Shop," where Fernández crystallizes his views on reinvention and underlines not only the comforting power of nostalgia but also its availability as a commodity when the store owner explains to his employee the value and purpose of selling old photographs: "They just want old-looking pictures. This is America. Everything is prefabricated, even the past."[29] This idea of nostalgia as a commodity and the manufacturing of cultural referents as depicted by Fernández in "Hologramas" and "The Pawn Shop" are exploited in the yearly Miami "flea market"/trade show/exhibit, *Cuba Nostalgia,* where vendors sell photographs, artifacts and cultural iconography—both authentic and reproductions—of Cuba's "Golden Age." The manufactured objects and memorabilia that blend with the few originals in these exhibits are simulacra, referring not to the reality of the past, but ultimately to the memories and recollections of the past. The event is described on their website (www.cubanostalgia.org) as "the premier Cuban event outside of Cuba, [. . .] a journey back in time for those who remember the island's glamorous times—and for those who never experienced them." The new generation, "those who never experienced [it]," is allowed a privileged glimpse into a past that is remembered through a spectacle that engages the five senses and allows for the confusion between the real and the illusion—a simulacrum. This historical recreation and manipulation of reality also evokes what Fredric Jameson, in his discussion of the representation of the past in literature and film, identifies as "not historical, but historicist . . . the random cannibalization of all the styles of the past, the play of random stylistic allusion. . . . This omnipresence of pastiche is not incompatible with a certain humor, however, nor is it innocent of all passion: it is at least compatible with addiction—with a whole historically original consumers' appetite for a world transformed into sheer images of itself and for pseudo events and 'spectacles' (the term of the situationists)."[29]

The fact that America is fertile ground for reinvention is a recurrent theme in Fernández's work and the author often situates his characters in the process of establishing new identities. As cultural sociologist Celia Lury states, "there is no natural one to one correspondence between the person, self identity and the individual."[30] This premise supports not only the possibility of personal reinvention and recreation through adopted and recreated identities but also that of the creation of altered, hyperreal environments that facilitate and accommodate perceptions of self-identity and of the elusive "American Dream."

The American Dream is as ephemeral as the exotic zoo in *Holy Radishes!* and while some of Fernández's characters are successful in their quest for this bountiful and generous state—Quinn and Mima from *Raining Backwards*, come to mind—most fail in their attempt to fully find their place in a perpetually uncooperative milieu. While his parodic presentation could easily be dismissed as a mere hyperbolic diversion, the author succeeds in capturing disillusionment and the transparency of incongruent affective, social and cultural realities. For those individuals without anchoring identities, reinvention is vital in the rebuilding of cultural and social frameworks. New realities are constantly being created and as Eco points out in his reflection on Ripley's museums, "[e]verything looks real, and therefore it is real; in any case the fact that it seems real is real, and the thing is real even if, like *Alice in Wonderland*, it never existed."[31] The Cuban exiles struggle to recreate and maintain their realities—their determined quest for a Paradisial recovery—and in the process, they create and sustain their own museum.

Notes

1. The first wave of Cuban exiles, also known as Historical Exiles, refers to those Cubans who came to the United States between 1960 and 1973 (the end of the Cuban Airlift) fleeing the Castro regime. All references to Cuban exiles in this paper refer to this particular group.

2. An exception might be Jewish-American literature where the predominant themes of assimilation and religious identity take precedent over victimization. There is also a recurrent comparison between American Jews and first wave Cuban Americans based on social class and connections to a diasporic brain drain. The term "Jews of the Caribbean" is sometimes used metaphorically to describe this first-wave of Cuban exiles and establish a parallel between the two diasporas with an emphasis on their common political, social and economic success in the United States.

3. Jean Baudrillard, *Simulacra and Simulation* (Ann Arbor: The University of Michigan Press, 1994), 3.

4. Ibid., 3.

5. Arlene Guerrero-Watanabe alludes to Baudrillard's idea of simulacrum while discussing Mirta's recreation scene in her article "Memory and Desire in Exile: The Narrative Strategies of *Raining Backwards.*"

6. Baudrillard, *Simulacra and Simulation*, 7.

7. Ibid., 1.

8. Whereas a few operatic mad scenes were written for male voices in the baroque period (Monteverdi to Handel), it was usually the heroine who got the more elaborate arias in the Romantic era.

9. Baudrillard, *Simulacra and Simulation*, 2.

10. "Kon-Tiki" has also appeared in Cuba in *La Gazeta de Cuba* (Sept-Oct 2004) and again in an collection of Fernández's works, *Entre dos aguas*, many for the first time in Spanish adaptations or transpositions. This short story also exists in an unpublished English version, "The Augustflower." While there are significant differences between the Spanish and English versions—"Kon-Tiki" contains lexical variations in character speech to establish their various places of origin, underline the fabrication of new identities, and enhance the outlandish elements of imitation—all quotations in this essay are from "The Augustflower," in order to use the author's original language.

11. The "wet-foot dry-foot policy" is an accord written in 1995 during the Clinton administration, as a revision of the Cuban Adjustment Act of 1966 stating that if Cubans touch U.S. soil, they are allowed to stay in the United States; however, if they are intercepted at sea, they will be returned to Cuba or to a third country. Those who make it to U.S. soil would qualify would qualify to apply for "legal permanent resident" status and eventually for U.S. citizenship.

12. From "The Augustflower," the unpublished, English version of "Kon-Tiki." See note 10.

13. Mirta's recreation of Varadero Beach has been discussed by several critics including William Deaver, Jorge Febles, Henry Pérez and Mary Vásquez. Mirta is not only an iconic figure who captures the essence of the first-wave Cuban exile community and their cultural disconnect; she reappears in several of Fernández's works as if to remind the reader of her significance: that in the incongruous environment where these characters exist, subsistence is only possible through simulacra, nostalgic revisionism, fantasy and reinvention. In this passage, Mirta has invited her friends to join her at a beach party she is hosting. The event, which no one attends, actually takes place in the character's bathroom and in her imagination: a private, confined space where recreation is possible, where pathos overshadows the ridiculous and the outlandish. After going to the local market and buying the necessary products, Mirta recreates Varadero, the Cuban beach, as she wants to remember it—an idyllic, tropical environment:

> Mirta was finishing pasting the last strip of the wall-paper with coconut trees and sunrises . . . she began to grind the cat litter into fine dust . . . she covered her bathtub with different depths of litter dust. Then she opened the faucet and started applying the blue-green dye to the water. To create the waves and the foam, she positioned the fan and put it on high and at the same time she dropped in four Alka Seltzer tablets. She then remembered that the breezes were always warm, so she placed the portable heater on top of the toilet . . . Mirta was happy. . . . She looked around the beach and felt lonely. She reached for the cabinet drawers and grabbed a handful of tiny plastic people which she lay throughout the beach in different positions. (*Raining Backwards* 55-56)

14. Jalena Sesnic, *From Shadow to Presence: Representations of Ethnicity in Contemporary American Literature* (Netherlands: Rodopi, 2007), 195.

15. In *Raining Backwards*, Mirta enchants her teenaged neighbor, Eloy, with fantastic tales about Cuba in exchange for erotic stimulation. This episode establishes memory, nostalgia and recreation as a commodity:

> In all the beaches in Cuba the sand was made out of grated silver, though in Varadero it was also mixed with diamond dust. . . . Could you tell me some more, Miss Mirta". . . . "If you stay a little longer I'll tell you about Foquie, the seal with the bright green eyes. Stay, stay . . . "and the white seals came around the Cape of Horne only because they had heard about the pleasures of Varadero, the most beautiful beach in the world." (10, 12-13)

16. Umberto Eco, *Travels in Hyperreality* (San Diego: Harcourt Brace, 1986), 7.

17. Roberto G. Fernández, *Holy Radishes!* (Houston: Arte Público Press, 1995), 54.

18. Ibid., 46.

19. Ibid., 57.

20. Fernández's allegorical frameworks are rich in symbolism, water being not only one of the most recurrent symbols but also the most powerful. The author uses the ambivalent symbolism of water as creator and destroyer of dreams and (hyper)realities. In the deluge scene in *Holy Radishes!*, the author seems to evoke a cleansing of the earth by God and a return to an ontological essence—not unlike the destructive water environment in *Kon-Tiki*. Water, however, is also used as a petri dish for fantasies and nostalgic recreation. See Mirta's recreation of Varadero in her tub (*Raining Backwards*), Varadero Beach as a Mount Olympus for a Cuban nostalgic mythology *(Raining Backwards)* and the chapter "Gulf Stream" from *En la Ocho y la Doce* where a boy—in a story inspired by the Elian Gonzalez episode—is saved from drowning by fantastic creatures to become a new Messiah. I am currently developing into an article the importance of water symbolism in the works of Fernández.

21. Roberto G. Fernández, *Holy Radishes!*, 63.

22. In "Operatic Transposition and the Romantic Aesthetic" I describe this escapist space in the context of a nineteenth century operatic mad scene as a return to "a time where love and joy dominated [the heroine's] world—a strong antithesis to her current state of emotional turmoil and deep anguish. In this reactionary state, the heroine is the victim of a failed or unrequited love affair, a lover's betrayal or a forced disconnection from an idyllic world." David de Posada, "Operatic Transposition and the Romantic Aesthetic in the Works of Roberto G. Fernández," *Caribe: Revista de Cultura y Literatura* 11.2 (2010): 23-42.

23. Roberto G. Fernández, *Holy Radishes!* 63.

24. Vijay Agnew, *Diaspora, Memory, and Identity: A Search for Home* (Toronto: University of Toronto Press, 2005), 9.

25. Eco's judgmental position on realism and American society readily supports Fernández's thematic framework: "Holography could prosper only in America, a country obsessed with realism, where, if reconstruction is to be credible, it must be absolutely iconic, a perfect likeness, a 'real' copy of the reality being represented." Eco, *Travels*, 4.

26. It is interesting to note that in a recent interview with the author, Fernández underlined his unfamiliarity with Baudrillard and Eco's references to holograms.

27. Roberto G. Fernández, *En la ocho y la doce* (Boston: Houghton Mifflin, 2001), 180. Excerpt translated from the Spanish by D. de Posada.

28. Roberto G. Fernández, *Holy Radishes!*, 202.

29. Frederic Jameson. *Postmodernism, or, the Cultural Logic of Late Capitalism* (Durham, NC: Duke University Press, 1992), 18.

30. Celia Lury, *Prosthetic Culture: Photography, Memory and Identity* (London: Routledge, 1998), 1.

31. Eco, *Travels*, 16.

Chapter 18

Having It Both Ways: The Simulacral and the Aura of the Real in Geoff Nicholson's *Bleeding London*

Leslie Haynsworth

Stuart London, one of the protagonists of Geoff Nicholson's 1998 novel *Bleeding London*, is so entranced with the city of London that he at one point imagines the city proliferating itself across the globe:

> let's say you wanted a new metropolis; in that case, you'd manufacture a brand-new second London, a perfect replica, identical in every physical detail. Then you could set it up in New Zealand or Dubai or Namibia, move in a population, leave it for a year or two, and then go back and see how much the new London had diverged from the old one.[1]

Stuart recognizes that these simulacral Londons can never entirely replicate the original, "real" London. "London isn't simply its architecture and hardware," he acknowledges. So

> The new London wouldn't for example, be a financial centre, wouldn't . . . have any history, wouldn't contain the same ethnic or social mix as the original. The climate would be different. But I still think it sounds like fun. I still think it might be better than starting from scratch every time someone wants to build something new.[2]

By the end of the novel, Stuart's fantasy is on the cusp of being fulfilled—but not quite in the way he imagined. Stuart and his wife Anita own a company that offers walking tours of London, and while he's been busy both dreaming of new Londons and trying to absorb as much of the real London as any person can, Anita has found them some new Japanese business partners. "They tell me the Japanese really love London," she explains, "but there are a couple of problems with it. First, it's too big, dirty, dangerous and expensive, and second, and rather more crucially, it's too far away from Japan."[3] The solution they've devised is

> a version of London out there in Japan. . . . All the tourist attractions of London will be there in one manageable site. There'll be Buckingham Palace, St. Paul's Cathedral, Big Ben, all scaled down to about half-size and all within a couple minutes walk of each other. . . . [P]eople want to shop when they come to London, so there'll be versions of Harrods and Selfridges. . . . There'll be walking tours, of course. There could be re-enactments of the Great Fire, the Plague, maybe even the Blitz, although the Japanese are a bit funny about the war, obviously.[4]

Confronted with this imminent realization of his vision, Stuart reacts not with joy but with hysteria, going fetal and "making a wet throaty sound [Anita] had never heard before," so that "she couldn't tell whether he was laughing or sobbing."[5]

Stuart's response points to both the central problem and the central promise of the simulacrum: it destabilizes both the identity and the meaning of the original, thereby allowing whatever value or prestige that adheres to the original to be co-opted in ways that aren't possible in relation to the original itself. Confronted with the actual city of London, there's nothing—or at least very little—that the Japanese can do to bend the city toward what they desire it to be. Their vision of a simulacral London, on the other hand, gives them the best of both worlds: everything they value most about the original London with none of its downsides or inconveniences.

Of course, what's lost to those who engage with even the most faithful replica of the original London is, as Stuart realizes from the start, the history, the climate, the people—the full complement of factors that make the real London most fully itself. Without these, the simulacrum tells a false and diminished story of what London is. Anyone who knows London only through the simulacrum knows only a partial and distorted London, a hollow shell in comparison to the full-bodied complexity of the original.

Bleeding London repeatedly compels its characters to navigate this vexed relationship between the simulacral and the real, and to parse what they can and can't get out of their engagements with each. While encounters with the real are always going to be richer and more meaningful, the novel nevertheless allows its characters to (at least sometimes) revel in the fact that encounters with the simulacral give us more experiential and interpretive control. But because there's no middle ground between the simulacral and the real—a place or an object is either the original or it's not—there's also no room to negotiate between the outcomes of our encounters with each. We can't have it both ways, can't simultaneously enjoy both the full value of the real and the greater control we exercise over the simulacral. We have to choose, one or the other. Or do we?

Bleeding London never suggests that the simulacral and the real can actually be blended into any kind of best-of-both-worlds combination. But the novel does propose that our interpretive practices with respect to each can allow us to *perceive* ourselves as having it both ways, to experience the simulacral as if it were the real even when we know it isn't. Essentially, what various characters in *Bleeding London* do with varying degrees of success is transmute their understanding of the aura of the real onto their encounters with the simulacral, or gain interpretive mastery over the real by reading it through their previous encounters with the simulacral. Via such interpretive practices, they experience themselves as having at least to some degree attained that middle ground whereupon they are neither overmastered by the interpretive tyranny of the real nor denied a rich and meaningful experience by the hollow sham nature of the simulacrum.

Walter Benjamin's concept of aura as the unique emanation of the real is a useful way into understanding the particular brand of postmodern mental hijinks performed by Nicholson's characters. In Benjamin's famous formulation, reproduction of any object inevitably dilutes or disrupts the object's aura. "The authenticity of a thing," Benjamin explains, "is the essence of all that is transmissible from its beginning, ranging from its substantive duration to its testimony to the history which it has experienced."[6] In other words, any object (or place) that is culturally and/or historically significant is going to radiate that significance out in the form of an aura, which is experienced by all who gaze upon the object with any kind of understanding of its history and/or its significance.

This, Benjamin argues, is why authenticity matters, why gazing upon an actual symbolically potent object is always going to be different than gazing at even a near-perfect simulacrum of it. Most crucially, reproduction alters the relationship between the object and its context: "technical reproduction," for example, "can put the copy of the original into situations which would be out of reach for the original itself. Above all, it enables the original to meet the beholder halfway."[7]

When the object's place and identity are stable, thoroughly grounded in the context of its actual existence, then it's the beholder who has to traverse all the space between the object and herself. Once the object is pulled out of its context, once the beholder feels authorized to allow a simulacrum of the object to stand in for the real thing, then the beholder begins to have more capacity to determine the meaning of the object and to control the relationship between it and herself. Rather than being a fixed signifier that bespeaks a very specific history, and a collectively determined set of meanings, the object becomes a floating signifier that, while still in some ways connected to that history and those meanings, is capable of being appropriated in other ways and for other ends. Is capable, for example, of having its potency as a symbol of national identity co-opted in the service of stirring up erotic desire, as when, in *Bleeding London*, a stripper named Gabby gets creative with Big Ben: "Shoving the plaster model of Big

Ben into herself was another little extra, but she had to be careful with it. That point on the top looked as though it could be lethal."[8]

For Benjamin, then, the capacity to reproduce symbolically potent objects is at once liberating—insofar as we are no longer such passive consumers of the meanings disseminated by these objects—and unsettling—insofar as these objects' capacity to serve as cultural touchstones, as repositories of stable and collective meaning, is diminished. Once Big Ben has been shoved up a stripper's vagina, no one who has witnessed that performance will ever see the real thing the same way again. And even the fact that we all know reproductions of the object are out there and that we can appropriate them as we choose means that the power the original once exerted over us is loosened. One could say, then, that in terms of the ways they mediate our relationship with the world, original objects and their simulacra exist in dialectical opposition to one another: real objects radiate stable meaning while simulacra proliferate indeterminacy; and in our engagement with the real, we are comparatively disempowered—the object tells us what it is, what it means, and what to do with it—while through our engagement with simulacra we gain increased authority over the meaning and uses of both the simulacra and the original object itself. Certain benefits accrue to us through our encounters with each, then. The real provides a reassuring degree of stability, certainty, and cultural authority. The simulacral provides more personal authority, more latitude to make meaning as we see fit. But there are certain downsides to each too: the real circumscribes our interpretive practices; the simulacral debases the real, making it less potent, less valuable as a cultural touchstone or a source of either collective or personal meaning.

And of course the more symbolically potent an object seems, the more likely it is to be reproduced. It's because Big Ben functions in the collective consciousness as a symbol that can stand in for all of London that so many people want a reproduction of it. Ironically, then, the stronger the aura of the original is, the more aggressively it's likely to be reproduced and reappropriated. At least in theory, then, the reproduction and cooptation of potent objects is radically democratizing, in that it functions to eradicate hierarchies of cultural value.

In that respect, Gabby's use of Big Ben in her strip act is politically one and the same with Stuart's efforts to remap the city of London by systematically walking each of its streets: "This would not," he proposes, "be anybody's particular version of London, not Dickens', not Pepys', not Betjeman's, and certainly not his own. Since every street was to be walked down, every street would have equal value and importance. He sensed something profoundly egalitarian in all this, a belief in a sort of democracy."[9] Stuart wants to perform a psychic leveling, to make every part of the city neither more nor less resonant than any other part. And there is, he feels, a political purpose to his plan: "Perhaps then his walking would be an act of reclamation, a way of taming the city, a way of saying that London was open and accessible to everyone, that it held no secrets, no unknowable horrors."[10] What Stuart wants, in effect, is to erase the city's aura, to release us all from the hold the city exerts, through its ability to radiate symbolic meaning, on all of us.

As the proprietor of a company that offers walking tours of London, however, Stuart is actively involved in helping to preserve the city's aura (or at least a simulacrum thereof that is mistaken by most people for the real thing). As he explains to one of his new tour guides, "The people who come on these tours, they're here to see sights, the Tower of London, Big Ben, Nelson's Column, and in one sense, of course, they've already seen them. They've seen them in books and on postcards and on television, and most importantly they've seen them in their mind. What they're usually doing when they come on a tour is having those images reinforced, making sure the Tower of London is the way they always imagined it. And in most cases it will be."[11]

What passes as, or is experienced as, aura is actually then, Stuart proposes, sort of a slippery collusion between the object and the viewer. The viewer wants to believe in aura, feels, as Benjamin suggests he should, as if the experience of gazing on the real object is more meaningful than the experience of encountering a simulacral version of the object. The viewer also accedes in some respects to what he believes is a consolidated, collective cultural interpretation of the object's meaning and value: Big Ben stands for London and the history and traditions thereof, and we all know at least some of the same things about all that. But in the end, the most important version of what Big Ben means is the preexisting version in the viewer's own mind, Big Ben and all its resonances refracted through the viewer's consciousness. And because it has, through its capacity to be reproduced, lost much of its capacity to signify the same things to everyone who sees it, Big Ben has to accede to the viewer's imposition of his own interpretations upon it.

All of which makes London itself sort of slippery for people in ways they sometimes struggle to understand. Big Ben is still Big Ben, and it still has an aura, but the potency of that aura is unstable, dependent on the degree to which the individual viewer understands it in its authentic formation and on the extent to which the viewer is willing to acquiesce to the aura's version of what Big Ben actually means. Yet, like the walkers on Stuart's tours, characters in the novel frequently evince a desire and a belief that aura is intact, that symbolically significant objects can function as stable and potent touchstones of meaning. Mick, who's down in London from Sheffield, is, for example, distressed by his first sight of Big Ben: "No. Get away. Surely Big Ben ought to be bigger than that. I have to say I'm a bit disappointed. I was expecting more. Maybe I was expecting too much."[12] A moment later, Mick expresses frustration with the geographic diffusion of London's most notable landmarks: "That's inconvenient, isn't it? You'd think they'd put all the famous places next to each other, wouldn't you?"[13] Ostensibly, the joke here is on Mick: it's as if he thinks all the "famous places" were manufactured expressly for tourist consumption. But in fact he's also calling for exactly what all the tourists who visit Big Ben, etc., really want: a palpable and highly concentrated sense of aura that produces interpretive clarity and certainty. Mick's vision is of a London whose touchstones are so close together that they function synergistically to produce a strong and consolidated narrative of what London is and what it means.

That, of course, is not, as Mick's response to the actual sight of Big Ben suggests, what the characters in the novel actually experience. Having become habituated to experiencing the city and its landmarks via simulacra—via the films they watch on TV, and the plaster models they see in gift shops—people have lost the capacity to respond to aura as they want to. The novel proposes that we've gotten so accustomed to exercising our capacity to do what simulacra allow us to do—meet the object in the middle—that we're incapable of recognizing aura for what it is. We all laugh at the mini-London Anita wants to construct in Japan—it seems absurdly simulacral, to the point where you can't imagine that anyone could even think they could possibly extract any meaning from an encounter with it. But that Japanese London is in fact just a logical extension of the way Stuart proposes we all interact with the actual London: we use it to consolidate the version of the city that already exists in our own minds.

The problem with aura in *Bleeding London* is thus not precisely its absence but rather its diminishment, which is caused by our inability to occupy the kind of passive subject position with respect to the object that would allow us to experience aura in its full potency. And it is primarily our habituation to simulacra, to an ersatz substitute that stands in symbolically for the real object, that makes us the wrong kinds of subjects to imbibe aura the way we might want to. Mick expects Big Ben to radiate tremendous amounts of meaning at him, but its capacity to do so has already been radically undercut by his repeated exposure to images of it elsewhere.

Bleeding London deliberately locates us in a world that's replete with simulacra, and where most characters evince a high degree of awareness of the simulacrarity of much of their experience, of, that is to say, the degree to which they are deliberately (or at least out of convenience) making use of simulacra as substitutes for the real thing. The consequence of this practice is that the meaning of everything is determined far more by the viewer/consumer than by the thing itself. Gone in this context, it would seem, is the capacity of an object, an institution, or even a city to signify clearly, via its aura, any kind of significant or consistent collective meaning. What we want to experience, or what we want to be true thus can come to take interpretive precedence over what we actually do experience or observe, as when, for example, Stuart and his employee Judy have "not really all that satisfactory or comfortable sex, but to a large extent it was the thought that counted, and the thought was pretty terrific."[14] If you desire for the sex to be good, you can make it so, even as you know that it really wasn't.

And yet, to some extent, the aura of culturally resonant objects or places still does have the capacity to stimulate a response. As Mick wanders around London, "Sometimes he'd find himself in a place he'd heard of, a famous place . . . and he'd feel simultaneously disoriented and at home. He couldn't quite square the fact that these places, which to his eyes looked so ordinary, so workaday, also carried such a weight of history or fame with them, and yet he felt good to find himself using such famous streets in his daily wanderings."[15] What Mick feels here is a vestigial sense of Benjamin's aura, an inchoate awareness of what he is supposed to feel in response to these famous places, and a vague sense of the prestige of these places and how that prestige might adhere to him via close contact with the actual place. What he experiences here, then, might be described as a simulacrum of aura—a concept which, in Benjamin's terms, is fundamentally oxymoronic and structurally impossible. But in a world where the proliferation of simulacra allows us to gain increasing interpretive mastery over the objects we encounter, there is in fact no reason why we can't will aura back into existence, why we can't, in fact, have it both ways, enjoying at one and the same time in the same encounter both the simulacrum's capacity to be appropriated as we see fit and the original object's capacity to tell us coherent and meaningful things about our culture and our history.

Given that it identifies nearly all of our experiences as being at least to some degree simulacral, and as being rooted more in our consciousnesses than in a true meeting-in-the-middle-type engagement between ourselves and the resonant objects or entities we encounter, *Bleeding London* seems less concerned with positing simulacrarity than with parsing it. The novel proposes a fairly complex taxonomy of simulacra and a similarly complex semiotics of responses to them. There is, on the one hand, the issue of just how simulacral a thing or an experience is; and on the other hand how aware of the simulacrarity of it the individual is, and how concerned the individual is with the fact of this simulacrarity. Stuart's father, for example, utterly confuses the simulacrum for the real when, in the midst of the Blitz, he sees a pit filled with the severed heads of waxwork figures from Mme. Tussaud's and, believing them to be the blitz-bombed heads of actual Londoners, faints in horror. At the other end of the spectrum of awareness is Stuart's sense that he lives in an age that's so simulacral, it inhibits his ability to produce meaningful art: "Stuart hoped his diary could be a fraction as worthy as Pepys', but sometimes he felt disadvantaged . . . You might say that these were plague years, but Aids [sic] had nothing on the Great Plague of 1665. . . . Stuart had the not uncommon sense that he was living a life that was simultaneously grotesque yet bland, cruel yet denied the drama of pestilence, fire, war, invasion, blitz."[16]

Yet it's also Stuart who recognizes that even as we may be surrounded by an increasing proliferation of simulacra, each individual has the capacity to push past the simulacral to engage directly with the real, and, at the same time, to fabricate the real as if he had the same authority over it that he can exert over the simulacral. The most common simulacrum of a city is a map, and as Stuart nears the end of his project of walking the entire city of London, he realizes that

> Soon I will no longer have use for a map. Maps are euphemisms, clean, clear, self-explanatory substitutes for all the mess and mayhem, the clutter and ambivalence and blurring and intermeshing weft and warp of the real places they purport to describe. They're no longer necessary since I have created a new

> London, not one made out of stone and brick, tarmac and concrete, but a London created out of memory, imagination, and shoe leather. I have dreamed it. I have made my dreams come true.[17]

Stuart's London is a collusion between the actual city and his consciousness. It renders the map, which was his original intertext for the city, irrelevant, suggesting that even when our encounters with the real are mediated by the simulacral, we can still see around or through the simulacrum to experience the actuality of the real. Stuart's London is also, however, not quite the same thing as London qua itself. It is London as *he* sees it, as *he* knows it, and, in fact, as he wants it to be. He experiences it as having an aura, a rich and complex aura which the map can't begin to express. But this aura itself is something of a fabrication. Certainly, as a co-creation of Stuart and the city, it's not Walter Benjamin's aura.

Nevertheless, Stuart does have it both ways here: he gets to experience what feels like the aura of the real and he gets to have the London of his own dreams. Likewise, when Mick in his wanderings around the city imbibes the aura of London even as he essentially has to will that aura into existence through his interpretive practice, he too is having it both ways: confronting the real and bending it toward what he wants or believes it to be. Clearly, the novel proposes, this approach can be taken too far: in Stuart's mind at least, the Japanese model that pushes the city as fully as possible toward what the Japanese want it to be is a case in point. But equally misguided, the novel suggests, is the belief in any sort of pure, unmediated engagement with the real, any full and true experience of aura. The characters in *Bleeding London* inhabit a murkily indeterminate space wherein they have to negotiate their own way between the simulacral and what they can find of the real. It's an uncertain, unstable, and at times perplexing space, but it's also, via its very uncertainties and confusions, a space that grants the individual enormous opportunities to reconfigure, reappropriate, or even reimagine both the simulacral and the real—to, in fact, have it almost any way we choose.

Notes

1. Geoff Nicholson, *Bleeding London* (Woodstock, NY: Overlook Press, 1998), 192. Copyright © 1997 Geoff Nicholson, published in 1997 by The Overlook Press, Peter Mayer Publishers, Inc. All rights reserved.
2. Ibid., 192.
3. Ibid., 332.
4. Ibid., 332-333.
5. Ibid., 333.
6. Walter Benjamin, "The Work of Art in the Age of Mechanical Reproduction," in *Illuminations* (NY: Shocken, 1969), 221.
7. Ibid., 220.
8. Nicholson, 117.

9. Ibid., 78.
10. Ibid., 77-78.
11. Ibid., 126-127.
12. Ibid., 50.
13. Ibid., 50.
14. Ibid., 130.
15. Ibid., 135.
16. Ibid., 247-248.
17. Ibid., 29.

Chapter 19

"There's really nothing left to explore": Imagineering, Innoventions, and Figments in Florida Studies

Lisa K. Perdigao

In Peter Weir's film *The Truman Show* (1998), a young Truman Burbank tells his teacher, "I would like to be an explorer, like the great Magellan" and his teacher replies, "Oh, you're too late! There's really nothing left to explore!" While Truman grows up in the artificial world of Seahaven, actors dissuade and prevent Truman from leaving it by "manufactur[ing] ways to keep him on the island." Yet, as an adult, Truman begins to recognize the artificiality of his world and longs to escape it; his treasure chest hidden in the basement reveals a map of Fiji and evidences his unabated desire for exploration. When an adult Truman announces that he wants to go to Fiji, his friend Marlon asks, "Where the hell is Fiji? Near Florida?" Truman's description of the location of Fiji, facilitated by a golf ball acting as a globe (Epcot anyone?), displays his recognition that there are still places to explore and even discover for the first time. While Marlon is an actor whose role is to help maintain Truman's world, his question about Fiji's location—in proximity to Florida—offers me a starting point for this essay. As Florida studies and its subset Disney studies have become veritable centers of the field of cultural studies, the geography of the field has been rewritten so that new sites (even Fiji) can be measured by their proximity to Florida. As critics, we may accept that "there's really nothing left to explore" or, following Truman's lead, as I elect to do in this essay, reexamine and redefine the map of Florida studies in and for the twenty-first century.

While *The Truman Show* was filmed in Seaside, Florida, offering the planned community as a site for exploring the line between reality and fictionality, its foundation is more expansive as a site for Florida studies. Throughout the film, Truman's world is circumscribed by Christof, his god-like creator and creator of *The Truman Show*, who, ostensibly channeling Jean Baudrillard, tells his audience, "We accept the reality of the world with which we are presented." In Baudrillard's *Simulacra and Simulation* (1981), first- and second-order simulations blur the boundary between reality and representation.[1] In third-order simulation, the "model precedes the real," creating the hyperreal, a "world without a real origin."[2] For Baudrillard and critics after Baudrillard, this is Disney, or, to borrow Stephen Fjellman's term, this creates "Distory," a "Disneyesque rewriting of history" (Fernandez 236).[3] In *Vinyl Leaves*, Fjellman writes, "In many of its attractions, pieces of literary tales are retold as if they were the whole story."[4] As a result, "History . . . is disaggregated" and "Geography is mixed up."[5] Indeed, Fiji might not be near Florida, but, as EPCOT's World Showcase demonstrates, America is next to Japan. Yet while such critical work can be read as, well, critical of the site, it also generates meaning when read as a "bricolage writ large in which all sorts of decontextualized messages make some kind of sense."[6] Assembling the "bricolage writ large" of Disney studies allows for a kind of sense-making at the site and offers, in a sense, a "perfect model" of the kind of simulation at work in cultural studies.

As work in Florida studies and, particularly, Disney studies continually (if not always) returns to Baudrillard's analysis of Disneyland, it ostensibly creates its own hyperreal, a model of "all the entangled orders of simulacra."[7] Rather than being a play with illusions, Florida studies plays with allusions to the site as well as critical and fictional representations of the site. The map Baudrillard evokes at the beginning of *Simulacra and Simulation* can be read as representative of what is at work in Florida studies. Describing a map that "precedes the territory," a model of "a real without origin or reality,"[8] Baudrillard's map extends further than that in the Borges fable he describes. With so many works written on Florida and, more specifically, Disney,[9] it seems as if there is nothing left to explore, certainly nothing left to discover. Yet mapping the field yields interesting possibilities in tracing origins, a Frontierland, and the future of a field, a Tomorrowland. As work in cultural studies constructs additional narrative frames, layers of readings, it further disguises the "real," in this case, with Florida as the site and center of study. Representations of Florida in literature, television, film, and social histories exist in abundance, depicting Florida as pastoral paradise and postmodern nightmare, and the critical work in the field, as representations of representations, also exhibits Baudrillard's successive phases of the image; the result is that the field itself becomes its own pure simulacrum.

From realistic representations to fantastic recreations, depictions of Florida as lived site and as textualized reality read like Disney World's own divisions, creating and sustaining stories of American culture: a Frontierland, Adventureland, Main Street USA, Liberty Square, Fantasyland, and Tomorrowland.

Gary R. Mormino's *Land of Sunshine, State of Dreams* (2005) captures, even in its title, popular images of Florida. He writes, "From its founding as an imperial outpost to its modern identity as a tourist empire, Florida has evoked contrasting and compelling images of the sacred and profane: a Fountain of Youth and a Garden of Earthly Delights, a miasmic hellhole and scuzzy wasteland"; it is a "powerful symbol of renewal and regeneration" whose "dreamscape constantly shifts."[10] While popular sentiment renders Florida as being more focused on "tomorrow's possibilities than today's realities," in the decades after 1940, "Everyman's Florida" emerged as a result of "revolutions in health care, leisure, and technology; the affordability and popularity of air conditioning and air travel; the selling and manufacturing of dreams by a loosely connected group of businessmen, developers, and press corps; the realization that Florida was a place where workers, retirees, and pleasure seekers could enjoy and share paradise."[11] But, in the twenty-first century, as Mormino notes, four hurricanes in 2004 changed Florida, turning the Sunshine State into the "Plywood State, the State of Fatigue, and the National Disaster State."[12] Still, Florida remains a symbol of progress, where "the loss of grove and forest" is overwritten with "newly minted words coined to explain Florida's growth: megalopolis, exurbs, microburbs, and boomburbs."[13]

Yet the language of the megalopolis cannot erase the loss of grove and forest. This tension in Mormino's descriptions is part of the stories of Florida. Campbell McGrath's *Florida Poems* (2002) tells multiple versions of the Florida story, illustrating a shift from the pastoral to the postmodern. In the sequence *A City in the Clouds*, McGrath depicts the presence of the alligators, the birth of a golden age, the introduction of machines, the return of the alligators, and the inevitable end. McGrath's Florida is a city in the clouds, abstracted dream, as well as postmodern nightmare, under the reign of Mickey Mouse who is both consummate politician and religious icon. McGrath's poems lament the loss of culture in Florida while they ostensibly celebrate Florida culture as spectacular. Throughout *Florida Poems*, McGrath reconstructs versions of Florida history and culture, but, rather than apply a chronological order to the narrative, McGrath writes like Baudrillard's cartographer, mapping the state and obscuring origins.

In "The Florida Poem," McGrath retells the history of Florida, tracing its foundation in fragmented words and enjambed lines. Throughout the poem, he tries to identify what is at the source of Florida but struggles to define it, highlighting the possibilities and limitations of his language. He looks for assistance from classical and postmodern figures, calling to the Floridian muse, playing with the conceits and the traditions of poetry, of Florida. Using elevated and colloquial terms, he advertises Florida as, conversely, both natural site and commercialized dream. Reconstructing a story of the place that dates back to the first explorers, back to Juan Ponce de León and Hernando de Soto, he rewrites their stories and the story of Florida. He presents us with an unmappable territory, once real and now too hyperreal. As he struggles with his Florida poem, he considers his sources along with Florida's resources. His story of Florida is one inscribed with absent presences, decentered centers. Perhaps this too is a better way of conceiving Florida studies.

While the site of Disney may be read as presenting "Distory," studies of Florida history and culture reveal similar signs. According to Baudrillard, "History is our lost referential, that is to say our myth."[14] Pieces of the Florida story are retold as if they were the whole story, and history is disaggregated. Floristory emerges. Maps of Florida (if not the "real" Florida) abound. Describing how one navigates the map that Disney provides, "a normalized American version of culture," Ramona Fernandez writes, "Those who are excluded for any reason must regularly draw alternative maps, rewrite narratives to move from a place of invisibility to imaginative inclusion. This is precisely what they do everyday: live lives of complex negotiation within a simulated and distorted landscape."[15] Part of Florida studies is the story of invisibility and imaginative inclusion. Mormino's book is a comprehensive one at over four hundred pages. Yet as part of the Florida History and Culture Series published by the University Press of Florida, it bears a foreword that states that "the quantity and quality of the literature on Florida's distinctive heritage and character have not kept pace with the Sunshine State's enhanced status" and its aim to "remedy this situation"; his text is framed by an absence of history, a neglect of Florida history and culture. Combining this statement with the perception that Florida studies, and Disney studies, leaves "nothing left to explore" presents a picture of Florida studies that reflects its source, a "constantly shifting" dreamscape and "powerful symbol of renewal and regeneration."

While the differing perceptions of the state of Florida studies resist a cohesive narrative, representations of Florida in both a literary tradition and a developing critical tradition suggest a shared ground and metaphor. While the University of Florida Press seeks to build a veritable library of books on Florida's history, Disney studies constructs a version of "Distory" grounded in a library of its own design. Fernandez's argument is that EPCOT is a "giant library, a mnemonic device, a topos for the memorization of 'facts' and a site for a kind of inductive association that threatens to run away with itself."[16] Florida becomes a textual and textualized center, existing as both lived space and "mnemonic device." Extending these terms, Celebration, Florida, as a planned community and realized vision of Walt Disney's masterplan for EPCOT, exposes this kind of history-making and re-presentation in Heritage Hall with its declaration of its founding. Douglas Frantz and Catherine Collins's statement that "Michael Eisner's hyped references to Celebration as the embodiment of Walt Disney's original dream of a real community at Epcot fueled the myth that living in Celebration would be the closest thing to living at Disney World. In the beginning, Celebration was a town where believing replaced knowing"[17] shows how, in a Baudrillardian turn, the history is replaced by the myth as Walt Disney's idea of EPCOT becomes Celebration.

An advertisement for Celebration, Florida asks, "Have you ever wanted to be part of something bigger than yourself?," sounding like a line from *The Truman Show* spoken by Christof. Marlon wonders "Who hasn't wanted to be somebody?" Truman is somebody; the viewers constantly watch him, have his calendars on their walls, hold pillows with his image on them. Seahaven is "a nice place to live," as their license plates attest, perhaps preferable to the real world. According to Christof, the world outside Seahaven is just another simulation; he tells Truman, "There is no more truth out there than there is in the world I created for you." The link between the film's Seahaven and Celebration is a solid one; in their study of Celebration, in a chapter entitled "Truman Didn't Sleep Here," Frantz and Collins discuss how visitors often mistake Celebration for Seaside, Florida, where the film was actually shot.[18] Celebration is like a fictional town; marketing it as "a super-modern Disney town with an old-time feel" places it in that Frontierland, between Main Street USA and Tomorrowland.

Gloria Naylor's novel *Linden Hills* (1985) presents a community that appears similar to Celebration or at least lies in the Frontierland between Celebration and Walt Disney's original EPCOT. Linden Hills is an affluent black community, yet, seen through the eyes of the young poet Willie and his friend Lester, it becomes a version of Dante's Inferno, where the characters lose themselves in their quests for material success. While Celebration is marketed as the "destination your soul has been searching for," Lester fears "selling the mirror in your soul."[19] Willie and Lester, demystifying the idealized community, turn to a historian, Dr. Braithwaite, who compiled an eleven-volume history of Linden Hills for answers. When Braithwaite tells them that his studies became "the record of a people who are lost,"[20] Willie and Lester question why he doesn't let people know what is happening. Braithwaite repeats his assumed objective position, saying, "I know that I can only hope to record that knowledge, not rectify it."[21]

While Willie and Lester reveal the limitations of Braithwaite's work, Braithwaite's statements about his work as an historian are more problematic. He says, "There are no absolute truths, and the best historians know that."[22] Yet, although he appears to acknowledge the limitations of the historian's representation of the "truth," he represents his work as compiling the whole or real story: "You take the family records and then you must view them in light of material from other sources . . . It becomes a whole series of checks and balances from numerous sources before you come up with the whole story, the real story if you will."[23] As Willie and Lester reveal, Braithwaite's story is limited; it is a master narrative provided to him by Luther Nedeed, the community's founder, and by his own window that reveals only one perspective on the site.

When Willie and Lester compile their own account of Linden Hills, they reflect Fernandez's idea of the negotiation of Disney. Willie and Lester are characters that "draw alternative maps, rewrite narratives to move from a place of invisibility to imaginative inclusion" in the "distorted landscape" of Linden

Hills.[24] They seem to understand another version of the story, Meaghan Morris' sense that "a history must assume that centers are actively transformed by their 'users'" and that "the history itself would count as one such transformation by a user."[25] Braithwaite's library offers Willie and Lester an education based on a "memorization of 'facts'" and "inductive association." Reading EPCOT, Fernandez cautions, "This is a library that relies on simulacra to create its narratives and soma to capture its audience while rigorously suppressing race, gender, and class."[26]

As Naylor's *Linden Hills* exposes the story of a place that depends on simulacra to create its narratives, it presents the library as the repository of those ideas. While Linden Hills is revealed to be less than ideal, Gary Ross' film *Pleasantville* (1998) reveals all that is pleasant in a community; Pleasantville in 1958 reads as part *The Truman Show* and part Celebration. While David knows the plots of *Pleasantville* and the emplotedness of the town, Jennifer despairs over her predicament. When the siblings first go to school in Pleasantville, they learn that its geography is isolationist and constrictive; in response to Mary Sue's question about what is outside of Pleasantville, the teacher, seeming to cross over from *The Truman Show*, says, "You should know the answer to that! The end of Main Street is just the beginning again." The library in Pleasantville tells a similar story: its books are blank. As David and Jennifer teach the residents of Pleasantville different ways to navigate their worlds, their world changes. David negates the teacher's lesson, telling others that "There are some places that the road doesn't go in a circle. There are some places where the road keeps going." He remembers stories he had read, and the pages of the books are filled in. Students flock to the library to learn more from the books about the world outside Pleasantville. In *The Truman Show,* the library is another set that Truman must escape to spend time with Lauren/Sylvia. Her books are signs that fail to generate meaning; she doesn't respond to the character's name listed on her book and does not understand the Japanese that Truman speaks and that she is supposedly studying. Yet, like the residents of Pleasantville, Truman draws alternative maps and redefines his space, ultimately escaping and rediscovering Sylvia at the film's end. This is the space of Florida studies, and cultural studies, redrafted, redesigned.

As *The Truman Show*, *Pleasantville*, and *Linden Hills* represent the impossibility of a perfect world and dystopic revision of the American dream, the story of Walt Disney's "American Project" reveals a similar narrative. Mormino's representation of the Florida of late twentieth-century literature and journalism as "a lost utopia—a dystopian, overdeveloped, land overrun by corporate theme parks, rapacious developers, and crazed drug lords"[27] could be taken from Carl Hiaasen's pages. In *Team Rodent* (1998), Hiaasen explains that in Haymarket, Virginia, in 1993, Disney attempted to construct a "three-thousand-acre, $650 million development," an amusement park with a history theme.[28] As the park was to be located six miles from the Civil War memorial, the Manassas National Battlefield, site of the battles of Bull Run, it met opposition and ultimately failed; Hiaasen writes, "The outcome proved genuinely heroic, though not in the way CEO Eisner foresaw."[29] It is intriguing that this failure, a "genuinely heroic" one, is staged at a historical site. According to Hiaasen, "Under the Eisner reign, nothing in the real world cannot be copied and refined in the name of entertainment, and no place is safe."[30]

And no place in Florida is safe from Hiaasen's critique. In *Native Tongue* (1991), Hiaasen creates a rival to Disney—the "Amazing Kingdom of Thrills"—and a plot to exact revenge on the Mouse whose "sprawling self-contained empire sucked tourists' pockets inside out."[31] At the beginning of the novel, after the Whelper family witnesses the violent death of the blue-tongued mango vole, they rush to the Amazing Kingdom of Thrills only to find a chaotic crime scene at the park. At the scene, Gerry Whelper says to her husband, "'Maybe they're filming a movie . . . Maybe it's not real'" (10). After the narrator reports, "But it was,"[32] the Whelper daughter, Jennifer, says, "'I'm so scared—can we drive to Epcot instead?'"[33] Terry Whelper answers, "'Excellent idea'"[34] and ushers everyone back to the car, back to Disney. In his writing in nonfiction and in fiction, Hiaasen continually returns to Disney, reconfiguring representations of Florida and redefining the map of Florida studies.

And here I also return to origins. While I discussed the significance of representations of Florida, I left leaves unturned. Fjellman's *Vinyl Leaves* suggests a model for understanding the hyperreality of Disney, Disney studies, and cultural studies. While the leaves are a natural foundation for the site and text (leaves on trees, leaves as pages), the phrase "vinyl leaves" conjures different meanings. Fjellman states that the Swiss Family Island Treehouse in Disney World's Adventureland shows how "art and history are transformed by dilution," as here "we have a fake tree holding a fake treehouse, representing a fake story told in a different medium from, but alluding to, a classic piece of literature, in an amusement park visited by 30 million people a year, most of whom are, like myself, enchanted."[35] Yet Fjellman does not despair over these vinyl leaves, at Disney, and those of his own writing. Instead, he says, "I love it! I could live there! I love its infinitude, its theater, its dadaism. I love its food, its craft, its simulations. It gets me to think, to remember, and to make up new fantasies . . . I crave its contradictions."[36]

Florida studies re-present signs of Florida culture, providing maps with marked paths and the potentiality of alternative routes. Between the vinyl and the leaves, nature and artifice transform the site as well as its art and history. Towards the end of *Native Tongue*, Molly McNamara, Mother of Wilderness, says, upon witnessing the destruction of the Amazing Kingdom, "Nature is a wonder. . . . Such power to renew, or to destroy. It's an awesome paradox."[37] Baudrillard writes, "We require a visible past, a visible continuum, a visible myth of origin, which reassures us of our end,"[38] yet, as we see in the example of Florida studies, a visible past is evasive. Instead, we have McGrath's *Florida Poems* that, instead of reassuring us, reveals that there is no visible myth of origin. Yet, rather than despairing over the loss of grove and forest, we can celebrate the power to renew, to take the vinyl leaves and find meaning. This bricolage that is Florida studies allows exploration, circumnavigation, reinterpretation, and reinscription. In the land of the hyperreal, in the space of Florida studies, denied access to a real story of origins, we retain the sense of both history and myth, discovering, at the source, the opportunity to begin again.

Notes

1. Richard Lane, *Jean Baudrillard*, 2nd ed. (NY: Routledge, 2009), 84.

2. Lane, *Jean Baudrillard*, 84.

3. Ramona Fernandez, "Pachucko Mickey," in *From Mouse to Mermaid: The Politics of Film, Gender, and Culture*, ed. Elizabeth Bell, Lynda Haas, and Laura Sells (Bloomington: Indiana UP, 1995), 236. Recent work in Disney studies define and redefine the construction of Disney ideologies. In their introduction to *From Mouse to Mermaid*, Elizabeth Bell, Lynda Haas, and Laura Sells differentiate five Disneys: Walt Disney the creator; Disney the studio; Disney the canon of popular film; Disney the multinational corporation; and Disney the ideology, "'a sign whose mythology and cultural capital is dependent on and imbricated in all the above manifestations of the name 'Disney'" (2). Compare this to Michael Real's use of the term "Disney universe" in his 1973 book *Mass-Mediated Culture*, a term that Janet Wasko employs to frame *Understanding Disney* (2001). Wasko represents Real's argument as extending from practicality, as the Disney organization used it, to evocativeness, as Disney products are "universal," to a more theoretical claim, for "the Disney message created 'an identifiable universe of semiotic meaning'" (3). Identifying the semiotic system of (and that is) Disney, Fjellman writes, "WDW literature is the retirement home of the inflationary descriptive adjective. We are ordered about by language that is constantly performative and directive. We are drained of interpretive autonomy. The rhetoric is so overexuberant that we stop noticing its intent: It is just normal Disneytalk" (13). For Fjellman, the performance of language offers a different ideology when it becomes "more sinister" with the "corporate-sponsored descriptions of the world, its history, and its future at EPCOT center" (13). Fjellman writes that the semantics of speech differ between the Disney sites—"nostalgia and fantasy at the Magic Kingdom, travel and culture at the World Showcase, science and technology at Future World" (14)—but the message is the same. According to Fernandez, "Disney 'reality' has permeated our cultural aesthetic—or perhaps it is more accurate to say that Disney reflects our aesthetic back at us" (236).

4. Stephen Fjellman, *Vinyl Leaves: Walt Disney World and America* (Boulder: Westview, 1992), 31.

5. Fjellman, *Vinyl Leaves*, 31.

6. Fjellman, *Vinyl Leaves*, 32.

7. Jean Baudrillard, *Simulacra and Simulation*, trans. Sheila Faria Glaser (Ann Arbor: U of Michigan Press, 2006), 12.

8. Baudrillard, *Simulacra and Simulation*, 1.

9. Wasko cites in her 2001 book *Understanding Disney* that her amazon.com search yielded 2,922 books with "Disney" in the title (4); my search in 2010 yielded 34,765 books. I echo her note that many are Disney products as opposed to Disney studies.

10. Gary R. Mormino, *Land of Sunshine, State of Dreams: A Social History of Modern Florida*. 2005 (Gainesville: U of Florida P, 2008), 2.

11. Mormino, *Land of Sunshine, State of Dreams*, 355-356.
12. Mormino, *Land of Sunshine, State of Dreams*, 359.
13. Mormino, *Land of Sunshine, State of Dreams*, 367. Mormino writes, "Tourism has inflicted a cruel and ironic dialectic upon Florida: the more people that come to enjoy lakes, river, and beaches, the less appealing and natural the places become" (119).
14. Baudrillard, *Simulacra and Simulation*, 43.
15. Fernandez, "Pachucko Mickey," 248.
16. Fernandez, "Pachucko Mickey," 237.
17. Douglas Frantz and Catherine Collins, *Celebration, U. S. A.: Living in Disney's Brave New Town* (NY: Holt, 1999), 314.
18. Frantz and Collins tell the story of a woman and her four children who stop Cathy for directions; one of the children asks, "Can you tell me where Truman lives?" (307). Cathy's response, "He was just a character. It was all make-believe. And, anyway, it wasn't filmed here. It was filmed over in Seaside, which is on the panhandle" (307) represents another Baudrillardian moment. Celebration becomes the fiction, Truman the reality.
19. Gloria Naylor, *Linden Hills*, 1985 (NY: Penguin, 1986), 59.
20. Naylor, *Linden Hills*, 261.
21. Naylor, *Linden Hills*, 262.
22. Naylor, *Linden Hills*, 262.
23. Naylor, *Linden Hills*, 263.
24. Fernandez, "Pachucko Mickey," 248.
25. Meaghan Morris, "Things to Do with Shopping Centers," in *The Cultural Studies Reader*, 2nd ed., ed. Simon During (NY: Routledge, 1999), 399, 400.
26. Fernandez, "Pachucko Mickey," 237. Fernandez extends the argument further, writing, "The bodies of people of color and women, and therefore the narratives surrounding them, are largely absent in EPCOT's displays" and "their agency is bleached away" (238).
27. Mormino, *Land of Sunshine, State of Dreams*, 9.
28. Carl Hiaasen, *Team Rodent: How Disney Devours the World* (NY: Random House, 1998), 21.
29. Hiaasen, *Team Rodent*, 22.
30. Hiaasen, *Team Rodent*, 79.
31. Carl Hiaasen, *Native Tongue* (NY: Vision, 1991), 46.
32. Hiaasen, *Native Tongue*, 10.
33. Hiaasen, *Native Tongue*, 13.
34. Hiaasen, *Native Tongue*, 13.
35. Fjellman, *Vinyl Leaves*, 2.
36. Fjellman, *Vinyl Leaves*, 16.
37. Hiaasen, *Native Tongue*, 472.
38. Baudrillard, *Simulacra and Simulation*, 10.

Selected Bibliography

Adema, Pauline. "Vicarious Consumption: Food, Television and the Ambiguity of Modernity." *Journal of American & Comparative Cultures* 23, no. 3 (2000): 113-24.

Agnew, Vijay. *Diaspora, Memory, and Identity: A Search for Home.* Toronto: University of Toronto Press, 2005.

Althusser, Louis. "Ideology and Ideological State Apparatuses (Notes Towards an Investigation)" (1970). In *Lenin and Philosophy and other Essays,* trans. Ben Brewster (1971), 121-176. Accessed June 19, 2010, http://www.marxists.org/reference/archive/althusser/1970/ideology.htm

Apter, Emily. "Technics of the Subject: The Avatar-Drive." *Postmodern Culture* 18.2 (Jan. 2008): n. pag. http://muse.jhu.edu/.

Auslander, Philip. *Liveness: Performance in a Mediatized Culture.* New York, NY: Routledge, 1999.

Avatar. DVD. Dir., James Cameron. Century City, Los Angeles, CA: Twentieth Century Fox, 2009.

Barthes, Roland. "La Mort de l'auteur." In *Le Bruissement de la langue*, 61–67. Paris: Seuil, 1984.

———. "Le Mythe, aujourd'hui." In *Mythologies*, 213–267. Paris: Seuil, 1957.

Baudrillard, Jean. *America.* Trans., Chris Turner. London: Verso, 1988.

———. *The Intelligence of Evil or the Lucidity Pact.* Trans. Chris Turner. New York: Berg, 2005.

———. *Simulacra and Simulation.* Ann Arbor, MI: The University of Michigan Press, 1994.

———. *Symbolic Exchange and Death.* Trans. Iain Hamilton Grant. Thousand Oaks, CA: Sage, 1993.

———. *The Vital Illusion,* edited by Julia Witwer. New York: Columbia University Press, 2000.

———. "Xerox and Infinity." In *The Transparency of Evil: Essays on Extreme Phenomena*, trans., James Benedict. London: Verso, 1993. Accessed June 25, 2010. http://insomnia.ac/essays/xerox_and_infinity/.

Bell, Elizabeth, Lynda Haas, and Laura Sells, eds. *From Mouse to Mermaid: The Politics of Film, Gender, and Culture.* Bloomington: Indiana UP, 2005.

Best, Steven and Douglas Kellner. *Postmodern Theory*. New York: The Guilford Press, 1991.

Best, Steven and Douglas Kellner. *The Postmodern Turn*. New York: The Guilford Press, 1997.

Bolter, Jay David and Richard Grusin. *Remediation: Understanding New Media*. Cambridge: MIT Press, 2000.

Boorstin, Daniel J. *The Image or What Happened to the American Dream*. New York: Atheneum, 1962.

Brummett, Barry. *The World and How We Describe It: Rhetorics of Reality, Representation, Simulation*. Westport, CT: Praeger, 2003.

Brunsdon, Charlotte. "Feminism, Postfeminism, Martha, Martha, and Nigella." *Cinema Journal* 44, no. 2 (2005): 110-16.

Buffardi, Laura E. and W. Keith Campbell. "Narcissism and Social Networking Web Sites." *Personality and Social Psychology Bulletin* 34 (2008): 1303-14, http://psp.sagepub.com/.

Butler, Judith. *Gender Trouble*. New York, NY: Routledge, 1990.

Caillois, Roger. *Man, Play, and Games*. New York, NY: Schocken Books, 1961.

Churchill, Ward. *On the Justice of Roosting Chickens*. (Oakland: AK Press, 2003).

Cole, Tim. *Selling The Holocaust: From Auschwitz to Schindler, How History is Bought, Packaged, and Sold*. New York: Routledge, 1999.

Croteau, David and Williams Hoynes. *Media Society: Industries, Images and Audiences*. London: Pine Forge Press, 2003.

Cubitt, Sean. *Simulation and Social Theory*. London: Sage, 2001.

De Souza e Silva, Adriana. "From Cyber to Hybrid: Mobile Technologies as Interfaces of Hybrid Spaces." *The Cybercultures Reader*. 2nd ed. Eds. David Bell and Barbara M. Kennedy, 757-772. New York, NY: Routledge, 2007.

Dead Space. Video Game. Redwood City: EA Redwood Shores, 2008.

Debord, Guy. *The Society of the Spectacle*. New York, NY: Zone Books, 1994.

Derrida, Jacques. "The Double Session." In *Dissemination*, trans., Barbara Johnson, 173-286. Chicago: U of Chicago, 1981.

———. "Faith and Knowledge." In *Acts of Religion*, edited by Gil Anidjar, 40-101. New York: Routledge, 2002.

———. *Specters of Marx*. Trans. Peggy Kamuf. New York: Routledge, 1994.

———. *The Gift of Death*. Trans., David Wills. Chicago: University of Chicago Press, 1995.

Derrida, Jacques, and Bernard Stiegler. *Echographies of Television*. Trans., Jennifer Bajorek. Oxford: Polity, 2002.

Duranske, Benjamin. "The 3d Internet Will Change How We Live," *Wall Street Journal*, July 17, 2008, Eastern ed., A13.

Dyer, Richard. *Heavenly Bodies: Film Stars and Society*. New York: Macmillan Education Ltd, 2004.

Eco, Umberto. *Postille a Il nome della rosa*. Milano: Bompiani, 1984.

———. *Travels in Hyperreality*. San Diego: Harcourt Brace. 1986.

Farmville. Zynga Games. 2009.

Fernandez, Jean. "'Master's Made Away With': Servant Voices and Narrational Politics in *The Strange Case of Dr. Jekyll and Mr. Hyde*." *LIT* 15 (2004): 363-387.

Fernandez, Ramona. "Pachucko Mickey." In *From Mouse to Mermaid: The Politics of Film,Gender, and Culture*, edited by Elizabeth Bell, Lynda Haas, and Laura Sells, 1-20, 236-254. Bloomington: Indiana UP, 1995.

Fernández, Roberto G. *En la ocho y la doce*. Boston: Houghton Mifflin, 2001.
———. *Holy Radishes!*. Houston: Arte Público Press, 1995.
Fjellman, Stephen. *Vinyl Leaves: Walt Disney World and America*. Boulder: Westview, 1992.
Foucault, Michel. *The Birth of the Clinic: An Archaeology of Medical Perception*. Trans., A.M. Sheridan Smith. New York: Panthcon, 1973.
———. *Discipline and Punish: The Birth of the Prison*. Trans., Alan Sheridan. New York: Pantheon, 1977.
Frantz, Douglas, and Catherine Collins. *Celebration, U.S.A.: Living in Disney's Brave New Town*. NY: Holt, 1999.
Freedman, Carl."England as Ideology: From 'Upstairs, Downstairs' to 'A Room with a View.'" *Cultural Critique* 17 (Winter 1990-1): 79-106.
Fuchs, Michael. "A Horrific Welcome to the Desert of the Real: Simulacra, Simulations, and Postmodern Horror." In *Landscapes of Postmodernity: Concepts and Paradigms of Critical Theory*, edited by Petra Eckhard, Michael Fuchs, and Walter W. Hölbling, 71– 90.Vienna: LIT Verlag, 2010.
Garrelts, Nate. *Digital Gameplay: Essays on the Nexus of Game and Gamer*. Ed. Nate Garrelts. Jefferson, NC: McFarland & Company, Inc, 2005.
Goldsmith, Kennneth, Reva Wolf, and Wayne Koestenbaum. *I'll Be Your Mirror: The Selected Andy Warhol Interviews, 1962-1987*. New York: Carroll and Graf Publishers, 2004.
Gould, Dana. "Homer the Moe." *The Simpsons*, season 13, episode 3, directed by Jen Kamerman, aired November 18, 2001. Los Angeles, CA: Twentieth Century Fox Home Entertainment, 2010. Blu-Ray.
Gubrium, Jaber F., and James A. Holstein. "Individual Agency, the Ordinary, and Postmodern Life." *The Sociological Quarterly* 36, no. 3 (1995): 555-70.
Heavy Rain. Video Game. Paris: Quantic Dream, 2010.
Heyes, Cressida J. "Cosmetic Surgery and the Televisual Makeover: A Foucauldian Feminist Reading," *Feminist Media Studies* 7.1 (2007): 17-32.
Hiaasen, Carl. *Native Tongue*. NY: Vision, 1991.
———. *Team Rodent: How Disney Devours the World*. NY: Random House, 1998.
Horton, Donald, and Richard R. Wohl. "Mass Communication and Para Social Interaction: Observations on Intimacy at a Distance." *Psychiatry* 19 (1956): 215-19.
Huizinga, John. *Homo Ludens: A Study of the Play Element in Culture*. Boston, MA: Beacon Press, 1950.
Jameson, Fredric. *Archaeologies of the Future: The Desire Called Utopia and Other Science Fictions*. New York: Verso, 2005.
Jameson, Fredric. *Postmodernism, or, The Cultural Logic of Late Capitalism*. North Carolina: Duke University Press, 2003.
Jenkins, Henry. *Convergence Culture: Where Old and New Media Collide*. New York, NY: New York University Press, 2006.
———. "Game Design as Narrative Architecture." Accessed July 4, 2010. http://web.mit.edu/cms/People/henry3/games&narrative.html
Jhabvala, Ruth Prawer. *The Remains of the Day*. DVD. Directed by James Ivory. Gloucester, England: Merchant Ivory Productions, 1993.
Johnson, Steven B. *Everything Bad Is Good for You: How Today's Popular Culture Is Actually Making Us Smarter*. New York, NY: Riverhead, 2005.
Juul, Jesper. *Half-Real: Video Games Between Real Rules and Fictional Worlds*. Cambridge, MA: The MIT Press, 2005.
Kamuf, Peggy, ed. "Introduction: Reading Between the Blinds." In *A Derrida Reader:*

Between the Blinds, xiii-xlii. New York: Columbia University Press, 1991.
Kaufman, Frederick. "Debbie Does Salad: The Food Network at the Frontiers of Pornography." *Harper's Magazine* (October 1, 2005): 55-60.
Keegan, Rebecca. *The Futurist*. New York: Crown, 2009.
Kelditty. "MTV Reality Meets for Drinks." In *The People We Love to Hate: Bitch Slapping Celebrities One Hit at a Time*. Accessed July 2, 2010. http://richfamousandgross.blogspot.com/2006_03_12_archive.html
Kellner, Douglas. *Baudrillard: A Critical Reader*. Oxford: Blackwell, 1994.
Kendrick, James. "Marxist Overtones in Three Films by James Cameron." *Journal of Popular Film and Television* 27, no. 3 (Fall 1999): 36-44.
Kingsepp, Eva. "Fighting Hyperreality With Hyperreality: History and Death in World War II Digital Games." *Games and Culture* 2 (2007): 366–375.
Lane, Richard. *Jean Baudrillard*. 2nd ed. NY: Routledge, 2009.
Lavery, David and Marc Leverette. "Afterword: Rereading *Seinfeld* after *Curb Your Enthusiasm*." In *Seinfeld, Master of its Domain: Revisiting Television's Greatest Sitcom*, edited by David Lavery and Sara Lewis Dunne, 203-19. New York: Continuum, 2006.
Lindvall, Terrance R., and J. Matthew Melton. "Towards a Post-Modern Animated Discourse: Bakhtin, Intertextuality and the Cartoon Revival." In *A Reader in Animation Studies*, edited by Jayne Pilling, 203–220. London: John Libbey, 1997.
Lister, Martin, Jon Dovey, Seth Giddens, Iain Grant and Kieran Kelly. *New Media: A Critical Introduction*. 2nd ed., New York: Routledge, 2009.
Lury, Celia. *Prosthetic Culture: Photography, Memory and Identity*. London: Routledge, 1998.
Lyotard, Jean-Francois. *The Differand: Phrases in Dispute*. (Minneapolis: The University of Minnesota Press, 1988).
Marx, Karl. *Capital*. Trans., Eden and Cedar Paul. Vol. 1. New York: E.P. Dutton & Co Inc, 1957.
Mass Effect 2. Video Game. Edmonton: BioWare, 2010.
Massumi, Brian. *Parables For The Virtual: Movement, Affect, Sensation*. Durham: Duke UP, 2002.
The Matrix. DVD. Dir., Andrew and Lana Wachowski. Burbank, CA: Warner Bros, 1999.
McGrath, Campbell. *Florida Poems*. NY: Ecco, 2002.
McLuhan, Marshall. *Understanding Media: The Extensions of Man*. Cambridge, MA: The MIT Press, 1994. Print.
Meadows, Mark Stephen. *I, Avatar: The Culture and Consequences of Having a Second Life*. Berkeley, CA: New Riders, 2008.
Mennell, Stephen, *All Manner of Food: Eating and Taste in England and France from the Middle-Ages to the Present*. Oxford: Blackwell, 1995.
Miller, Vincent. *Consuming Religion: Christian Faith and Practice in a Consumer Culture*. New York: Continuum, 2005.
Mittell, Jason. "Narrative Complexity in Contemporary American Television." *The Velvet Light Trap* 58 (2006): 29–40.
Mormino, Gary R. *Land of Sunshine, State of Dreams: A Social History of Modern Florida*. 2005. Gainesville: U of Florida P, 2008.
Morris, Meaghan. "Things to Do with Shopping Centers." In *The Cultural Studies Reader*. 2nd ed. Ed., Simon During 392-409. NY: Routledge, 1999.

Nakamura, Lisa. *Digitizing Race: Visual Cultures of the Internet.* Minneapolis: U of Minnesota, 2008.

Nathanson, Elizabeth. "As Easy as Pie: Cooking Shows, Domestic Efficiency, and Postfeminist Temporality." *Television New Media* 10 (2009): 311-30.

Naylor, Gloria. *Linden Hills.* 1985. NY: Penguin, 1986.

Newbury, Michael. "Celebrity Watching." *American Literary History.* 12.½ (Spring Summer 2000): 272-283.

Newitz, Annalee. "When Will White People Stop Making Movies Like 'Avatar'?." *i09,* December 2009. http://io9.com/5422666/when-will-white-people-stop-making-movies-like-avatar

Nietzsche, Friedrich. "Die fröhliche Wissenschaft." In *Friedrich Nietzsche: Werke in drei Bänden,* vol. 2, edited by Karl Schlechta, 7-274. Munich: Hanser, 1959.

Okrent, Daniel. "An Introduction to Rotisserie League Baseball." In *Rotisserie League Baseball,* edited by Glen Waggoner, 3-7. New York: Bantam, 1984.

Orwell, George. *The Orwell Reader: Fictions, Essays, and Reportage by George Orwell.* New York: Harcourt, Brace and Company, 1956.

Park, Robert E. "The Natural History of the Newspaper." *The American Journal of Sociology* 29:3 (Nov. 1923): 273-289.

Parillo, Wayne. "Why I Love Paris Hilton" in *Behind the Curtain.* Accessed April 9, 2007. http://wparillo.blogspot.com/2007/01/why-i-love-paris-hilton.html

The Passion of the Christ. DVD. Dir. Mel Gibson. Perf. Jim Caviezel. Warner Home Video, 2004.

Pérez-Firmat, Gustavo. *Life on the Hyphen: The Cuban-American Way.* Austin: University of Texas Press, 1994.

Peter, Tom A. "Study Abroad through Second Life," *Christian Science Monitor* (July 3, 2008): 25.

Plato. "The Republic, Book X." In *The Norton Anthology of Theory and Criticism,* edited by Vincent B. Leitch, William E. Cain, Laurie A. Finke, Barbara E. Johnson, John McGowan and Jeffrey J. Williams, 67-80. New York: Norton, 2001.

Pleasantville. DVD. Dir., Gary Rôss. New York: New Line, 2004.

Prensky, Marc. *Digital Games Based Learning.* New York: McGraw-Hill, 2001.

Quinn, Kevin G. *Sports and Their Fans: The History, Economics and Culture of the Relationship Between Spectator and Sport.* Jefferson, NC: McFarland, 2009.

Renzetti, Elizabeth. "James Cameron's Avatar: A symphony in blue and green." *Globe and Mail: Arts,* July 2010. http://www.theglobeandmail.com/news/arts/james-camerons-avatar-a-symphony-in-blue-and-green/article1405271/email/

Robbins, Bruce. *The Servant's Hand: English Fiction from Below.* Durham: Duke UP, 1993.

Rose, Frank. "Lonely Planet," *Wired* (August 2007): 141-44.

Rushkoff, Douglas. *Cyberia: Life in the trenches of hyperspace.* Manchester: Clinamen Press, 2002.

Sanders, Lise Shapiro. "Consuming Nigella." In *Feminism, Domesticity, and Popular Culture,* edited by Joanne Hollows Stacy Gillis, 151-163. New York: Routledge, 2009.

Schechner, Richard. *Performance Studies: An Introduction.* 2nd ed. New York, NY: Routledge, 2006.

Scherzinger, Karen. "The Butler in (the) Passage: The Liminal Narrative of Kazuo Ishiguro's *The Remains of the Day.*" *Liberator* 25.1 (April 2004): 1-21.

"Season 7 Invitation to the Set." HBO video. 2:47. http://www.hbo.com/curb-your-enthusiasm/about/video/invitation-to-set.html

Second Life. Accessed June 17, 2010. http://secondlife.com/

Serazio, Michael. "Virtual Sports Consumption, Authentic Brotherhood: The Reality of Fantasy Football." In *Sports Mania: Essays on Fandom and the Media in the 21st Century*, edited by Lawrence W. Hugenberg, Paul M. Haridakis, and Adam C. Earnheardt, 229-42. Jefferson, NC: McFarland, 2008.

Sesnic, Jalena. *From Shadow to Presence: Representations of Ethnicity in Contemporary American Literature*. Netherlands: Rodopi, 2007.

The Simple Life. "Daycare." TV. Dir., Kasey Barrett, Alicia Bean, Ross Breitenbach, Katherine Brooks, Rich Kin, Laura Korkoian, Jeff Savenick and Troy VanderHayden, March 9, 2005. Los Angeles: 20th Century Fox Television, Bunim-Murray Productions.

Skirvin, Tim. "How To Post About Nazis and Get Away With It—The Godwin's Law FAQ." *Internet FAQ Archive*. Accessed February 15, 2009. http://www.faqs.org/-faqs/usenet/legends/godwin/.

Smith, M.W. *Reading Simulacra: Fatal Theories for Postmodernity*. Albany: State University of New York Press 2001.

South Park. MP3. "You Have 0 Friends." The Comedy Network, 2010.

Sparks, Glenn and Cheri Sparks. "Effects of Media Violence." *Media Effects: Advances in Theory and Research*. Eds. Jennings Bryant and Dolf Zillmann, 269-285. New York: Lawrence Erlbaum Associates, 2002.

Spath, Eberhard. "P.G. Wodehouse's Jeeves: The Butler as Superman." *Functions of Literature: Essays Presented to Erwin Hoff on his Sixtieth Birthday* 1 (1984): 269-81.

Steiger, Janet. *Media Reception Studies*. New York: New York University Press, 2005.

Stiles, Kristine and Peter Selz, ed. *Theories and Documents of Contemporary Art: A Source Book of Artists' Writings*. California: University of California Press, 1996.

Strathausen, Carsten. "The Philosopher's Body: Derrida and Technology." *New Centennial Review* 9 no. 2 (2009): 139-64. http://ehis.ebscohost.com/

Strauss, Leo. *Natural Right and History*. Chicago: University of Chicago Press, 1953.

Swenson, Rebecca. "Domestic Divo? Televised Treatments of Masculinity, Femininity and Food." *Critical Studies in Media Communication* 26, no. 1 (2009): 36-53.

Terestchenko, Michael. "Servility and Destructiveness in Kazuo Ishiguro's *The Remains of the Day*." *Partial Answers* 5.1 (2007): 77-89.

Thomas, Angela. *Youth Online: Identity and Literacy in the Digital Age*. New York: Peter Lang, 2007.

Tom Clancy's Splinter Cell: Conviction. Video Game. Montreal: Ubisoft Montreal, 2010.

Trifonova, Temenuga. "Is There a Subject in Hyperreality?" *Postmodern Culture* 13, no. 3 (2003). Accessed March 15, 2010. doi: 10.1353/pmc.2003.0025

Trimm, Ryan S. "Inside Job: Professionalism and Postimperial Communities in *The Remains of the Day*." *LIT* 16 (2005): 125-61).

The Truman Show. DVD. Dir., Peter Weir. Hollywood, CA: Paramount, 1999.

Valentine, Gill. "Eating In: Home, Consumption, and Identity." *Sociology Review* 47, no. 3 (1999): 491-525.

Wasko, Janet. *Understanding Disney: The Manufacture of Fantasy*. Oxford: Blackwell, 2001.

Weide, Robert B. "Funny Business: Stranger than Fiction." *DGA Quarterly* Fall 2005. http://www.dga.org/news/dgaq_1105/beg_funnybiz-1105.php3?section=news

Weisman, Jordan. "42's 5 principles of arg." *CREATIVITY*. 2006. Accessed July 2, 2010. http://www.creativity-online.com/news/425-5-principles-of-arg/106069.

Westerman, Molly. "Is the Butler Home? Narrative and the Split Subject in *The Remains of the Day*." *Mosaic* 37.3 (2004): 157-70.
Young, Richard P. *Roots of Rebellion: The Evolution of Black Politics and Protest Since World War II*. (New York: Harper and Row, Publishers 1970).
Zizek, Slavoj. *The Plague of Fantasies*. London: Verso, 2008.
———. *Tarrying with the Negative: Kant, Hegel, and the Critique of Ideology*. Durham, NC: Duke UP, 1993.
———. *Welcome to the Desert of the Real!* London: Verso, 2002.

Index

About the Contributors

Amarnath Amarasingam is a doctoral candidate in the Laurier-Waterloo Ph.D. in Religious Studies, and holds a Social Sciences and Humanities Research Council of Canada Doctoral Fellowship. His research interests are in social movements, ethno-nationalism, radicalization, religion and science, and media studies. He is the editor of *Religion and the New Atheism: A Critical Appraisal* (Brill, 2010), and has published articles in *Studies in Religion*, the *Journal of Contemporary Religion*, *Canadian Ethnic Studies*, and *Muslim Minority Affairs*. He has presented papers at numerous national and international conferences, and has contributed to *The Huffington Post*, *The Daily Beast*, *The Toronto Star*, *The Province* (Vancouver), and *The Washington Post*'s *On Faith* blog. He is currently writing his dissertation entitled *Pain, Pride, and Politics: Sri Lankan Tamil Nationalism in Canada.*

Lian Amaris is a writer and artist living in New York City. She has master's degrees in Interactive Telecommunications and in Performance Studies, both from New York University, and has contributed articles on new media in/and the arts to *Theatre Journal, TDR: The Drama Review*, and *Explorations on Media Ecology,* along with several edited collections. After three years as a professor of performance studies and digital media, Amaris joined the Education Division of the Brooklyn Museum where she oversees programs for college and graduate students and works to bridge the gaps between performance, visual arts, and new media through public programs.

Matt Cornish is a Doctor of Fine Arts candidate at the Yale School of Drama, where he received his M.F.A. in Dramaturgy and Dramatic Criticism. He has worked as a teaching fellow at the Yale School of Drama and in the Theater Studies and English departments at Yale College. Currently a Fulbright Graduate Fellow at the Freie Universität Berlin, he is working on his dissertation: "Last Stop, Germania: Representing History and Nation in the Post-Reunification German Theater." His essays have been published in *Theater*, *PAJ*, and *Theater Journal*.

David de Posada is Associate Professor of French and Spanish at Georgia College & State University. He earned his Ph.D. in French Literature with a minor in Contemporary Spanish Peninsular Women Writers and U.S. Latino Writers from The Florida State University. He also holds graduate degrees from Florida International University (M.S., Modern Language Education) and from the Université de Reims Champagne-Ardenne, France (D.E.A. [Master of Advanced Studies], French Medieval and Renaissance Literature). His biographical-critical essay on Puerto Rican author Edgardo Vega Yunqué was included in *Latino and Latina Writers*, as part of Thomson-Gale's Scribner Writers Series. His most recent article, "Operatic Transposition and the Romantic Aesthetic," appears in *Caribe*, a journal of Caribbean literature and culture.

Robin DeRosa (Editor) has a B.A. from Brown and a Ph.D. from Tufts University. She is currently Associate Professor of English and Women's Studies at Plymouth State University in New Hampshire. Her most recent book is *The Making of Salem: the Witch Trials in History, Fiction and Tourism* (McFarland). She has also had work published in two recent collections, *American Indian Rhetorics of Survivance* (Pittsburgh UP) and *Women as Sites of Culture* (Ashgate), as well in several journals.

Naomi L. Fosher is a graduate student of English and Library & Information Studies at the University of Rhode Island. Her research interests include gender studies, the "cybrarian" (cyborg librarian) and reality/authenticity. She recently assisted in the editing of *End of Days: Essays on the Apocalypse from Antiquity to Modernity* (McFarland).

Michael Fuchs is Assistant Professor in the Department of American Studies, University of Graz (Austria). He is the author of *The Simpsons: Postmodernism, Postmodernity, and America's Most Animated Family* (VDM, 2008) and co-editor of *Landscapes of Postmodernity: Concepts and Paradigms of Critical Theory* (LIT Publishers, 2010). Currently, he is co-editing two books, *ConFiguring America: Iconic Figures, Visuality, and the American Identity* (Intellect Books) and *Time, Space, and Place: Constructing American Identities* (Graz UP), and finishing his Ph.D. thesis on self-reference in horror movies. Michael's research interests are eclectic, including video games, adult cinema, comics and graphic novels, postmodernist literature and drama, and non-linear systems.

Jenna Gerds is a Ph.D. candidate at Wayne State University in Detroit and has a Bachelor's degree from the University of Michigan. She has presented and written on American literature by authors ranging from John Dos Passos and Sinclair Lewis to Jeffrey Eugenides and Hubert Selby, Jr. Her study of literature is juxtaposed with an interdisciplinary examination of film, advertisement and television.

Benjamin D. Hagen is A.B.D. in the Department of English at the University of Rhode Island. He is currently at work on his dissertation, provisionally titled *Planes Over Chaos: Thinking in Deleuze, Woolf, Eliot, and Lawrence.* His most recent publication, "A Car, a Plane, and a Tower: Interrogating Public Images in *Mrs. Dalloway,*" appeared in the September 2009 issue of *Modernism/Modernity.*

Emily Hall has her B.A. and M.A. from Wesley College. She is currently a Ph.D. student who is studying Contemporary British Literature at University of North Carolina, Greensboro. She has spoken at several conferences, and has written on authors as diverse as Ayn Rand, Jeannette Winterson, and Gabriel Garcia Marquez.

Mihaela P Harper is a doctoral candidate (A.B.D.) in the Department of English at the University of Rhode Island. She is currently completing her dissertation, which brings together literary works by contemporary European and American writers and offers a new conception of the term *anomie.* Her B.A. is from the American University in Bulgaria and her M.S. is from Southern Connecticut State University. She has presented her work at the Modern Language Association and the Northeast Modern Language Association conventions, among others, and participated in the London Critical Theory School at Birkbeck Institute in 2010.

Leslie Haynsworth holds a Ph.D. in English from the University of Virginia and is pursuing an M.F.A. in creative writing at the University of South Carolina. She is co-author with David Toomey of *Amelia Earhart's Daughters* (William Morrow) and is currently co-editing a collection of essays, *Vexed by the Victorians: 21st-Century Encounters with 19th-Century Fiction,* with Maria LaMonaca. Her scholarship has been published in *Victorian Newsletter, Victorians Institute Journal, English Literature in Transition,* and *GenXegesis* (University of Wisconsin Press). Her stories, essays, and articles have appeared in *Fourth Genre, The Common Review, Roanoke Review, Gulf Stream, Marie Claire, The Denver Post, Publishers Weekly,* and elsewhere.

Randy Laist received his Ph.D. from the University of Connecticut and is currently Assistant Professor of English at Goodwin College. He is the author of *Technology and Postmodern Subjectivity in Don DeLillo's Novels* (Peter Lang) and the editor of *Looking for* Lost*: Critical Essays on the Enigmatic Series*

(McFarland). He has also published numerous articles on literature, film, and new media.

Amanda LeBlanc is a Ph.D. student at The University of Alberta (English and Film Studies). She has an M.A. in Cultural Studies and Critical Theory from McMaster University, where she was a Canada Graduate Scholar.

Cathie LeBlanc has a B.A. from Dartmouth College and a Ph.D. in Computer Science from Florida State University. She is currently Professor of Digital Media at Plymouth State University. She is the author of *Project-Based Software Engineering: An Object-Oriented Approach* (Addison Wesley Longman), as well as numerous articles for journals and anthologies.

Ann K. McClellan earned an A.B. from the University of Michigan and a Ph.D. from the University of Cincinnati. She is currently Associate Professor of English and the chair of the Women's Studies department at Plymouth State University. She is the author of *Mind Over Mother: Gender, Education, and Culture in 20th Century British Women's Fiction* (under contract), and she has published articles in several journals, including *English Literature in Transition, Women's Studies,* and *LIT:Literature, Interpretation, Theory.*

Lisa K. Perdigao has a Ph.D. from Northeastern University, and she is currently Associate Professor of English at the Florida Institute of Technology. She is the author of *From Modernist Entombment to Postmodernist Exhumation: Dead Bodies in Twentieth-Century American Fiction* (Ashgate 2010) and is co-editor, with Mark Pizzato, of *Death in American Texts and Performances: Corpses, Ghosts, and the Reanimated Dead* (Ashgate 2010). She has published articles in collections on twentieth-century poetry and fiction, and, most recently, in *Approaches to Teaching Faulkner's As I Lay Dying* (MLA).

Andrew J Ploeg is a Ph.D. candidate at the University of Rhode Island, currently working on his dissertation, entitled *Becoming Undecidable: Conceptualizing the Numinous Through Space, Movement, and Novelty.* His most recent article, "'I Will Remain Silence and Scream': Edmond Jabès and the Wound and Witness of Language," is scheduled for publication by *Shofar: An Interdisciplinary Journal of Jewish Studies* in Spring 2012.

Glenda Shaw-Garlock is finishing her Ph.D. in Communication at Simon Fraser University. She has had articles in two recent collections, *Essays on Monsters and the Monstrous* (Oxford) and *Readings in Advertising, Society, and Consumer Culture* (ME Sharpe), as well as in the *International Journal of Social Robotics* and the *Journal of Advertising.* Glenda has also been a researcher for the Council for Policy and Research in Science and Technology (CPROST) and is a faculty member at Capilano University, Vancouver, Canada. Glenda's areas of teaching and academic research include: research methods; the history of technology and society; affective technology and human users; the gendering of technological artifacts; as well as the representation of science and technology in film and popular culture.

Katheryn Wright has a Ph.D. from Florida State University, specializing in media studies and cultural geography. Her dissertation, *Into the Screenscape: Screens, Bodies, and the Biopolitics of the Population,* explores the formation and significance of the 'screenscape' in contemporary visual culture. She has work published in the online journal *Rhizomes,* and will have her chapter "Nine Inch Nails' *Year Zero* and the Biopolitics of Media Convergence" included in the forthcoming collection *The Politics of Post-9/11 Music: Sound, Trauma, and the Music Industry in the Time of Terror* (Ashgate 2011).

www.ingramcontent.com/pod-product-compliance
Ingram Content Group UK Ltd.
Pitfield, Milton Keynes, MK11 3LW, UK
UKHW020126290926
15471UKWH00003B/306

9 780739 184585